SPECIAL TOPICS IN ELECTROCHEMISTRY

SPECIAL TOPICS IN ELECTROCHEMISTRY

Edited by

PETER A. ROCK

Department of Chemistry,
University of California,
Davis, CA 95616, U.S.A.

ELSEVIER SCIENTIFIC PUBLISHING COMPANY
Amsterdam – Oxford – New York 1977

ELSEVIER SCIENTIFIC PUBLISHING COMPANY
335 Jan van Galenstraat
P.O. Box 211, Amsterdam, The Netherlands

Distributors for the United States and Canada:

ELSEVIER NORTH-HOLLAND INC.
52, Vanderbilt Avenue
New York, N.Y. 10017

Library of Congress Cataloging in Publication Data
Main entry under title:

Special topics in electrochemistry.

"Based for the most part on papers presented at the Symposium entitled 'Teaching of Electrochemistry' which was held on August 31, 1976 at the 172nd ACS Meeting in San Francisco, CA."
Bibliography: p.
1. Electrochemistry--Congresses. I. Rock, Peter A., 1939-
QD551.S65 541'.37 77-11122
ISBN 0-444-41627-7

Printed in The Netherlands

PREFACE

In the last decade the field of electrochemistry has undergone a rapid expansion. This expansion is a result of several factors, but the major ones are undoubtedly energy shortages, the desire for higher quality environments, and the greatly expanded efforts to understand bioelectric phenomena. Because of the rapid expansion of research efforts in electrochemistry, significant gaps have developed between the coverage of electrochemistry in existing chemistry textbooks and the research literature. This book represents an attempt to bridge these gaps in several of the subfields of contemporary electrochemistry. The level of presentation, and the extent of the coverage, of the various topics has been designed for (a) senior and first-year-graduate students who wish to survey a variety of the research areas of contemporary electrochemistry, and (b) for chemistry teachers who would like to update and expand the coverage of electrochemistry in their courses. The material in the book is based for the most part on papers presented at the Symposium entitled "Teaching of Electrochemistry" which was held on August 31, 1976 at the 172nd ACS Meeting in San Francisco, CA. The symposium was sponsored by the Division of Chemical Education, Inc.*

Special thanks go to Jacqueline Gutierrez for her assistance in the preparation of the camera-ready copy of the material.

University of California
Davis, California
May 26, 1977

Peter A. Rock

* Royalties from the sale of this book have been assigned by the contributors to the ACS Division of Chemical Education, Inc.

CONTENTS

ADVANCED ELECTROCHEMICAL ENERGY SYSTEMS

L. R. McCoy
Rockwell International Corporation, Atomics International Division

ABSTRACT

Vigorous activity is now underway to develop new electrochemical energy systems and to improve the performance of more familiar ones such as the lead-acid battery. This effort is attributable to a relatively recent awareness on the part of both industry and government of the need for better, cheaper, and mo: reliable systems of this type. In the case of industry, large potential markets for new products offer an incentive to pursue these investigations. In the case of government, the future economic health of the nation may be at issue. This awareness began with the oil embargo several years ago, which forced a reassessment of the present and future energy needs and resources of the United States. This paper examines the probable nature of future energy sources, the resultant need for improved electrochemical energy systems, the approaches being taken to satisfy these needs, and some of the problems to be overcome if these efforts are to be successful.

PRESENT AND FUTURE ENERGY SOURCES

Sources of energy used in the U. S. and their principal end applications in 1975 are summarized in Figure 1. Nealy half of the total energy used was supplied in the form of petroleum with the next largest fractions being supplied by natural gas and coal in that order. These three fossil fuels thus supplied 93% of the total energy consumed by this country in 1975. More than 40% of the petroleum used was imported, and this percentage has been steadily increasing for several decades [1]. U. S. production peaked in 1970 and is down about 20% since that time [2]. Natural gas production peaked in 1973 and has decreased about 10% in the last two years [2, 3].

The principal end products - heat, transportation, and electric power generation - are rather evenly split, but it should be noted that synthetic organic chemicals derived from fossil fuels are included in the industrial use. Transportation and electric power together comprise more than half

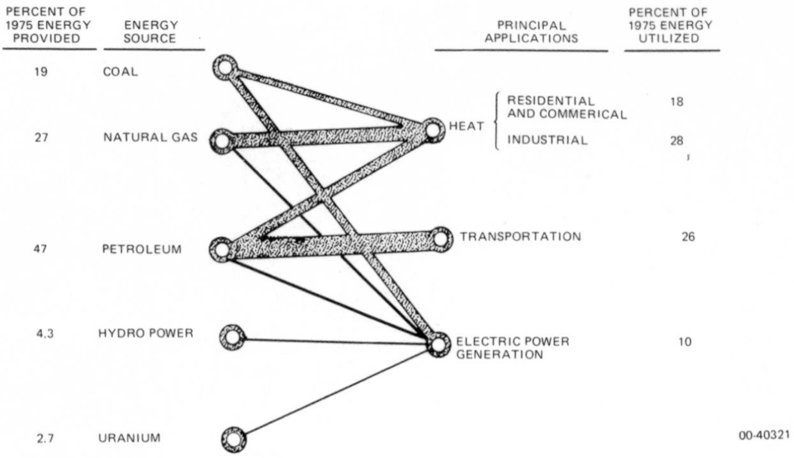

Figure 1. Present Day Significant Sources of Energy

of the total. While all fossil fuels are applicable to all end applications, other energy sources (hydro-power and nuclear fuel) can produce only electrical energy. The fact that fossil fuels are unique sources of carbon-hydrogen compounds useful for organic synthesis may make them too valuable to burn at some time in the future.

One of the energy sources shown in Figure 1, hydro-power, has a limited ability to expand to meet future needs. There remain few further opportunities in this country to harness large rivers for their energy. The percentage contribution of this source to the total must, therefore, diminish with time.

Figure 2 shows the proven domestic reserves of the fossil and mineral (uranium) fuels considered in Figure 1, together with their 1975 rates of consumption. On this basis, there exists a 350-year supply of coal, a 6-year supply of oil, an 11-year supply of natural gas, and either a 250- or 15,000-year supply of nuclear fuel if these were used at their indicated rates of consumption [1, 4, 5, 6]. In the case of nuclear fuel, the shorter time applies to present, conventional reactors while the longer time is that achievable if fast breeder reactors, which generate more new fuel than they consume, were available. These figures would be drastically revised if all of the energy were to be supplied by one fuel alone. On this basis, there would exist only a 71-year supply of coal, a 3-year supply of oil or gas, a 7-year supply of fuel for conventional nuclear reactors, but a 430-year supply of fuel for fast breeder reactors.

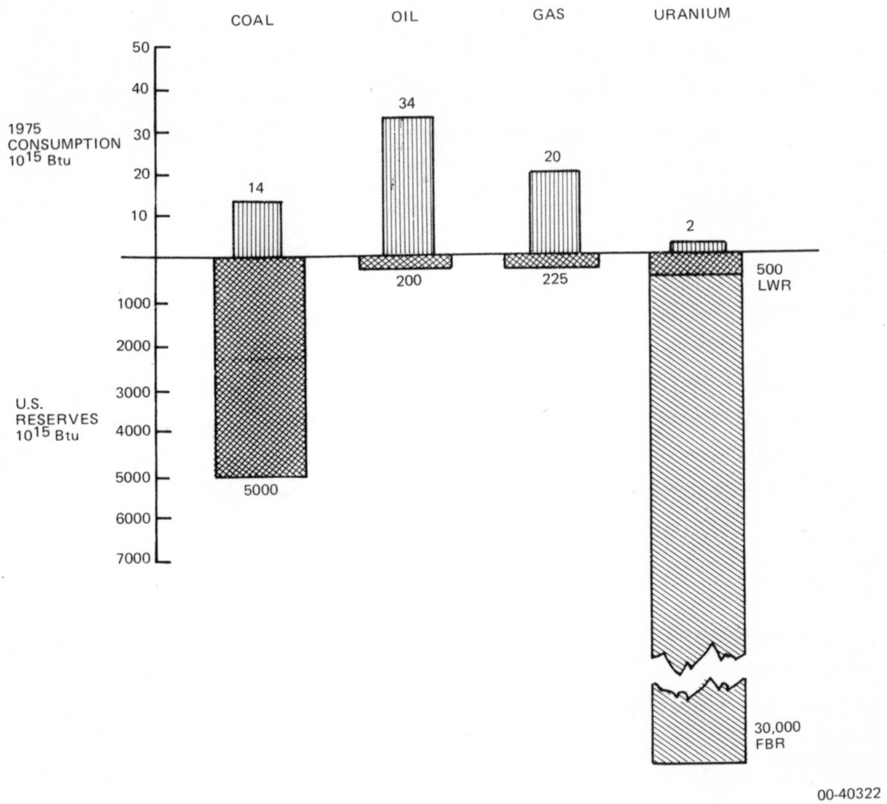

Figure 2. Proven Recoverable U. S. Reserves and 1975 Consumption

Two complications remain with the figures presented above. The proven reserves will be expanded by further exploration and may markedly extend the useful life of fossil fuels beyond that indicated by Figure 2. In the case of oil and gas, the need to seek these in such inhospitable regions as the arctic or at sea assures us, however, that they will be increasingly expensive. The second factor is the probable development of new sources of energy. Shale oil and tar sands are obvious examples, but economic and environmental problems indicate that these sources will yield more expensive fuels than those now being used. In the non-fossil fuel category, direct solar and fusion energy offer the hope of relatively non-polluting, large-scale energy sources. To a lesser degree, energy from winds, tides, ocean thermal gradients, and geothermal activity offer future promise. The problems in developing these energy sources vary in degree from the technically formidable in the case of fusion to essentially economic in the case of solar energy. Some, like geothermal energy, are severely restricted as to usable geologic sites within the scope of existing technology and are unlikely to contribute more than a small percentage of the total energy consumed. A similar situation exists in the case of tidal energy. All of the last named, however, have one point in common, they are essentially inexhaustible sources of energy.

THE NEED FOR ENERGY STORAGE SYSTEMS

The energy picture presented above indicates that an increasing share of energy will be supplied as electricity, and that conservation of fossil fuels is desirable if not essential. Most of the future sources of electrical energy will involve large capital costs [7, 8]. Whereas fuel costs and capital costs contribute roughly the same fraction to the cost of electricity from coal-fired generating plants, capital costs constitute a major fraction of the cost of electricity from nuclear plants. In the case of solar energy, capital costs will account for almost the total cost of the energy produced. Efficient use of plant capacity will, therefore, become increasingly important if costs of electricity are to be minimized. Unfortunately, as discussed below, power demand varies widely with the time of day and the period of the year. Satisfying peak demand requires that plant capacity go unused at other times. It is apparent that this situation would be greatly improved if electrical energy could be stored when demand was low, and then supplied to augment base capacity when demand was high. Some of the proposed energy sources, wind, solar, tides, are discontinuous energy sources, and some means of storing electrical energy is required if these are to supply power over a 24-hour period.

The fact that an increasing fraction of total energy will be available in the form of electricity suggests that energy consumption should be directed away from direct use of fossil fuels and toward this energy form. An estimate has been made that over 50% of petroleum consumption will be in the transportation sector by 1980 [8]. As almost three-fourths of this petroleum is used to fuel personal cars, it is apparent that electric cars could have a major impact on petroleum usage.

Two immediate and obvious energy storage needs, therefore, exist in the light of future energy sources: bulk storage of electrical energy by utilities and rechargeable batteries for electric vehicles.

Peak power demand varies with the time of day and also with the season, as shown on Figure 3 reprinted from a report issued by the Electric Power Research Institute [9]. The electric utility must provide service for the highest demand with some reserve beyond that. Peak demand is usually satisfied by bringing on line additional capacity from older and less fuel-efficient plants or, in more recent times, by gas turbine generators. The latter are low capital cost, short lead-time power sources when compared to base plant capacity, and the use of combined cycle units where exhaust gases make steam to drive turbines can be highly efficient [8].

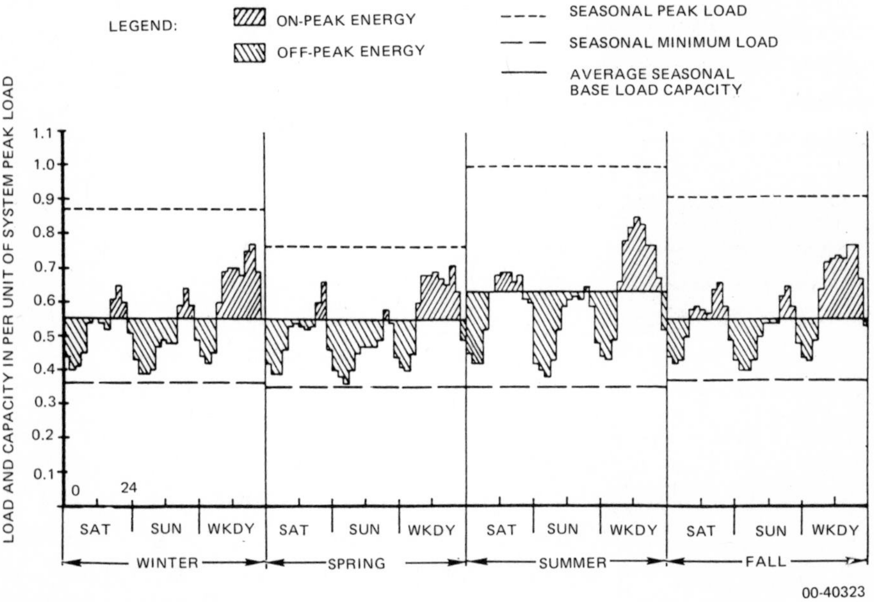

Figure 3. Distribution of Off-Peak and On-Peak Energy for Representative Electric System

This service could also be provided by electrochemical storage systems such as secondary batteries. When demand is low, the batteries can be charged and the stored energy can later be fed into the distribution system, permitting base capacity to be used at higher efficiency.

Electrochemical energy systems offer several advantages over gas turbines. They are noiseless, nonpolluting, and can be located in heavily-populated areas, reducing transmission costs. They are also potentially cheaper to operate if the cost of fuel continues to rise, as expected. These systems may take two forms, batteries and fuel cells. Although fuel cells are presently designed to use a variety of cracked hydrocarbon feed stocks, and thus will consume fossil fuels, they could at some time in the future also utilize hydrogen generated by electrolysis of water [10, 11]. For the short-term, at least, heavy emphasis is being placed on the development of rechargeable (secondary) batteries for utility use.

If the advantages of secondary batteries are to be realized, bulk storage of electricity must be economically as well as technically feasible. Peak power must be produced at a cost lower than that obtained from additions to base plant capacity and must compete with alternate means of supplying peak power needs. Here the situation is more complex, involving fuel cost and availability on a present and future basis, geographical location, environmental considerations, and system reliability [9].

The cost of an electrochemical energy storage system is represented only in part by that of the batteries. Interface equipment needed to convert ac line power to dc during charge, and to convert battery power from dc to ac during discharge, also contributes to the costs. Electrical and thermal controls required for the battery system also contribute both to capital and operating costs. The cost of using the energy storage system depends both on the useful life of the system and the duty it serves. For two systems with an equivalent installed cost, that with a longer life will produce power at a lower cost. It should be noted that although batteries have been referred to as <u>energy</u> storage systems, their value to the utility is calculated on a <u>power</u> output basis. Given two systems of equivalent energy storage, the system capable of delivering higher power, albiet for a shorter period of time, offers an economic advantage [7, 8, 9].

While the above discussion is concerned primarily with capital costs, operating costs are represented by a number of factors of which the energy efficiency is one of the more important. Defined as energy output divided by energy input, it is apparent that for a 50% overall efficiency, the cost of electricity delivered would, in the absence of other factors, be twice that of the energy supplied. Also important is the fact that the energy wasted per cycle will be dissipated in the form of heat which will introduce parasitic power losses incurred in the operation of pumps, blowers, etc. that are required to discharge this wasted energy to the atmosphere, thus decreasing the overall energy efficiency of the system. The above considerations have led to provisional economic and performance goals for secondary batteries for this application. These generally include an energy efficiency in excess of 70%, a life greater than 2,000 cycles, and a cost of around $20 to $30/kWh [7, 8, 9].

While the goals shown above are necessary, they are not sufficient for this application. Although most electrical systems can be regarded as nonpolluting, some of the systems now under study could, under some circumstances, release explosive or noxious gases. The highly active materials used in others could also, under extreme conditions, be potentially hazardous. Safety of operation, particularly if these systems are to be located in urban centers, is therefore an important consideration.

Finally, the large number of batteries required to make a meaningful contribution to electric utility energy storage on a nation-wide basis requires that sufficient quantities of materials be available to satisfy the needs for this application.

The quantity of materials involved can be illustrated by estimates that these storage systems could represent at an installed capacity of 2×10^{11} Wh by 1985 [7]. Approximate working voltages have been assumed for four secondary battery systems in Table 1. Nickel is common to three of these. The metals involved and the charged and discharged species are shown for each battery type. From the number of electrons transferred in each case and the working voltages, the tons of metal required have been calculated for the total storage requirement. It is apparent in the case of nickel that this varies with the voltage assumed for each battery system. The metal production for 1972, shown in Table 1, is based on information found in Reference 4. Finally, the active metal requirement is shown as a percentage of the 1972 production. It is apparent that the metal needs are significant in all cases except for iron and are enormous for cadmium.

This illustration is subject to many qualifications, as metal salts rather than the metal itself are used in many cases. The table does suggest, however, that the widespread use of these battery systems could have a major impact on the production of these metals if they were to be used in this application. It should be noted that these calculations have been made using the unlikely assumption of 100% utilization of the active material and ignores the fact that some metals such as lead and nickel are also used as structural materials. If these matters are taken into

TABLE 1

SECONDARY BATTERIES FOR ELECTRIC UTILITIES, RESOURCE REQUIREMENTS BASED ON 2×10^{11} Wh

Battery	Material	Voltage	Electrons	Wh/lb (1)	Tons $\times 10^6$ (1) Required	Tons $\times 10^6$ Produced 1972	Percent Requirement of 1972 Production
Nickel-Zinc	Nickel [NiOOH, $Ni(OH)_2$]	1.5	1	311	0.32	0.70 (2)	46
	Zinc (Zn metal, ZnO)	1.5	2	558	0.18	0.48 (3)	38
Lead-Acid	Lead (PbO_2, $PbSO_4$)	2.0	2	235	0.22		
	Lead (Pb, $PbSO_4$)	2.0	2	235	0.22		
					0.44	0.62 (3)	71
Nickel-Iron	Nickel [NiOOH, $Ni(OH_2)$]	1.1	1	228	0.44	0.70 (2)	63
	Iron [Fe, $Fe(OH)_2$]	1.1	2	479	0.21	89.0 (4)	0.2
Nickel-Cadmium	Nickel [NiOOH, $Ni(OH)_2$]	1.3	1	269	0.37	0.70 (2)	53
	Cadmium [Cd, $Cd(OH)_2$]	1.3	2	281	0.36	0.004 (5)	9,000

(1) As metal
(2) World production
(3) From domestic ores, U.S.
(4) As pig iron
(5) From other metal refining operations

account, the real requirements could exceed those shown here by factors of two to ten. As cadmium is obtained only as a byproduct of lead and zinc refining processes, its production is uniquely inelastic with respect to increased demand. Using an assumed cost of $25/kWh, an installed capacity of 2 x 10^{11} Wh would represent $5 billion worth of batteries, a not inconsiderable potential market for private industry.

ELECTRIC VEHICLE BATTERIES

Electric vehicles do not use fossil fuels, at least directly, do not contribute to urban smog, and do offer the possibility of an overall reduced energy consumption [8]. This saving in energy would result if the energy consumed in the production and distribution of gasoline were considered. As electric vehicles would be recharged overnight in most cases, their use would, in fact, result in some load leveling in urban centers [12].

The basic problems hindering widespread acceptance of electric vehicles today are their small size and limited range - 30 to 60 miles - depending on the conditions under which they are driven. Both problems stem from the inadequacies of lead-acid batteries presently used to power electric vehicles [13, 14, 15]. Range can only be extended by increasing the proportion of the total vehicle weight represented by the weight of the battery. As this is done, the structural weight of the vehicle must be increased to support the added weight. Weight from both sources requires additional power for acceleration, shortening the vehicle range, and the process is self-defeating. Obviously, the answer lies in better batteries containing more energy per unit weight and volume [16].

A major distinction between energy storage systems for vehicles and for utilities lies in the need for high energy density vehicle batteries. While energy density is important to utility storage systems to the extent that it affects land, building, and installation costs, it is a crucial matter for vehicle application. For this reason, it is likely that secondary batteries represent the most likely form of electrochemical energy storage systems for vehicles although fuel cells cannot be totally disregarded [17].

Although weight energy density is the factor most often considered in this application, volume energy density must also be considered. If a vehicle is to carry passengers, it is obvious that the batteries cannot occupy all available space. Further, vehicular operation requires short bursts of high power on acceleration, and specific power output (kW/kg) is no less important [18]. Usually (and particularly in the case of lead acid batteries), the energy density decreases as power density is increased. This relationship for a number of battery systems, present and projected, is shown in Figure 4 from a report prepared by the U. S. Department of Commerce [19]. It is apparent from this figure that values for energy density have little meaning unless the rate of discharge is also specified.

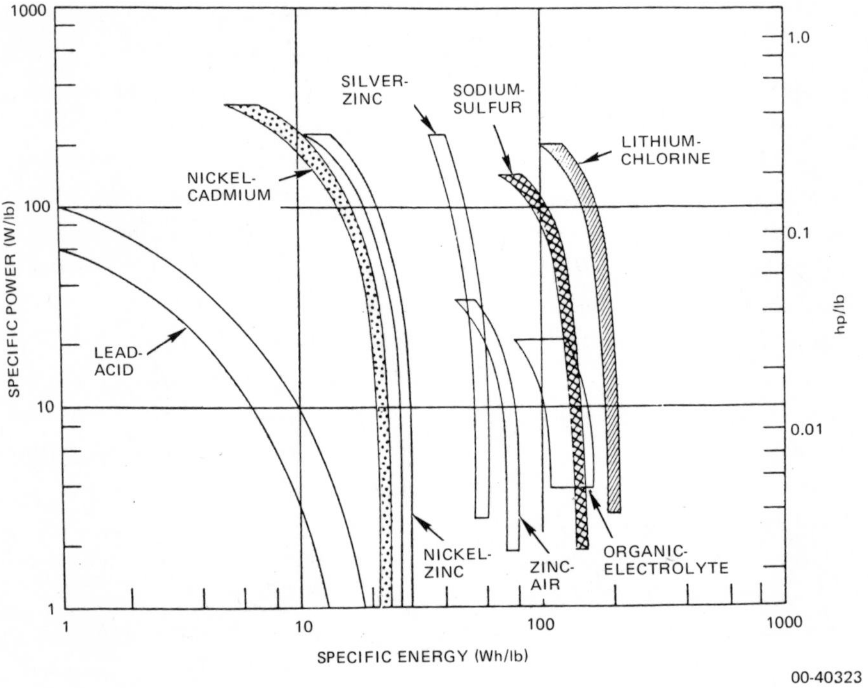

Figure 4. Specific Power vs Specific Energy for Batteries

As in the case of utility energy storage systems, cost and battery life are important. Using rough assumptions of a 50 kWh battery costing $50/kWh, this results in a battery cost of $2,500, not an insignificant fraction of the cost of the vehicle and an unacceptable one if frequent replacement were required. Again, the availability of the materials required for such a large-scale application is an important consideration.

The market for such batteries would be enormous if electric vehicles were to reach widespread acceptance. Assuming that one million electric vehicles were produced per year (about a tenth of the present car market) and using the above values of 50 kWh per vehicle and \$50/kWh, a \$2.5 billion market for batteries would exist each year.

STATUS OF PRESENT ELECTROCHEMICAL SYSTEMS

The situation with respect to existing secondary batteries for vehicle use is summarized in Table 2. This summary was presented by Landgrebe of the Energy Research and Development Administration at a workshop at Argonne National Laboratory (ANL) [13]. Only silver-zinc has a really high energy density but the high cost and short cycle life make it impractical for this application. Indeed, from a cost standpoint, only the lead-acid battery can be viewed as a serious candidate for vehicle use. Unfortunately, the lowest cost lead-acid battery has an unacceptably short cycle life. Longer cycle life can be achieved but at the expense of higher cost or a lower energy density. The situation with respect to load-leveling batteries is similar in that cost and/or limited life make the lead-acid battery marginal for this application [7].

TABLE 2

COMPARISON OF TODAY'S SECONDARY BATTERIES

Battery Type	Cost(1) (\$/kWh)	Energy Density By(2)		Life(3) (Cycles)
		Weight (Wh/kg)	Volume (kWh/ft^3)	
Silver-Zinc	900	120	8.8	100-300
Nickel-Cadmium	600	40	3.6	300-2000
Nickel-Iron	400	33	1.4	3000
Lead-Acid				
Motive Power	50	22	2.6	1500-2000
Submarine	80	28	2.0	400
Golf Car	35	35	2.2	300
Electric Vehicle	100	35	2.8	500-800

(1) Cost to the user

(2) Battery capacity is inversely related to rate of discharge. The values shown are for the 6-hour rate

(3) Cycle life depends on a number of factors, including depth of discharge, rate of charge and discharge, temperature, and amount of overcharge. Range shown is from most severe to modest duty

Although a variety of fuel cells have been built, largely with funds provided by Federal Agencies such as the National Aeronautics and Space Administration (NASA), none are known to be available as off-the-shelf, commercial items.

APPROACHES TO ADVANCED ELECTROCHEMICAL SYSTEMS

Several approaches exist for seeking improved systems. A high theoretical energy density can be achieved by selecting reactants with low equivalent weights and/or reactant couples with high theoretical voltages. The H_2-O_2 fuel cell (1.2 V) represents an extreme case of the use of low molecular weight reactants. Further, oxygen supplied by air need not be stored and does not contribute to the weight of the system. The sodium-sulfur couple uses low atomic weight reactants and the cell voltage is attractively high (2.1 V).

Theoretical energy density calculations are made using the open-circuit cell voltage and are not, therefore, representative of what can be achieved with a working cell or battery. Voltage losses are sustained under actual load as a result of a number of factors including electrode reaction kinetics (activation overpotential), concentration gradients (activity overpotentials), and internal cell resistance. Slow reaction kinetics can be speeded by increasing the temperatures at which the cell is operated. The fuel cell is an example where this approach has been followed to avoid the use of the expensive platinum catalyst required for ambient temperature operation. High temperatures may also be required to achieve low electrolyte resistivity.

Finally, the life, performance, and cost of more conventional systems can be improved by substitution of new and improved materials of construction, or by redesign to optimize components for specific applications. In this case, a higher fraction of the theoretical potential of the system is sought. This procedure is presently being used to upgrade lead-acid and nickel-iron batteries. To the extent that significant improvements may be expected from this approach, these batteries qualify accurately as advanced systems.

If these approaches offer promise of new and improved energy systems, they may also give rise at the same time to new problems. Thus, the use of highly reactive, low molecular weight materials involves compatibility problems with the electrolyte or structural components, high

TABLE 3

CLASSIFICATIONS OF ELECTROCHEMICAL ENERGY SYSTEMS

(Sheet 2 of 3)

	Cell Reactions (Electrolyte)	Voltage(5)	Theoretical Energy Density (Wh/kg)	Most Likely Application(6)
2. Redox Cells (example)	$3CrSO_4 + H_2CrO_4 + 3H_2SO_4 \xrightleftharpoons{(H_2SO_4)} 2Cr_2(SO_4)_3 + 4H_2O$	1.7	160	LL
3. Zinc-Halogen				
a. Chlorine Hydrate	$Zn + Cl_2 \xrightleftharpoons{(ZnCl_2)} ZnCl_2$	2.1	862	EV, LL
b. Organic Bromine Complex	$Zn + Br_2 \xrightleftharpoons{(ZnBr_2)} ZnBr_2$	1.8	381	EV
4. Zinc-Air	$Zn + \frac{1}{2}O_2 \xrightleftharpoons{(KOH)} ZnO$	1.6	1054	EV
II. High Temperature Systems				
A. Solid Electrolyte				
1. Sodium Negative Electrode				
a. Sodium-Sulfur	$2Na + 3S \xrightleftharpoons{(2)} Na_2S_3$	2.1	793	EV, LL
b. Sodium-Antimony Chloride	$3Na + SbCl_3 \xrightleftharpoons{(3)} 3NaCl + Sb$	3.0	1057	EV, LL

TABLE 3

CLASSIFICATIONS OF ELECTROCHEMICAL ENERGY SYSTEMS

(Sheet 3 of 3)

	Cell Reactions (Electrolyte)	Voltage(5)	Theoretical Energy Density (Wh/kg)	Most Likely Application(6)
2. Fuel Cells	$H_2 + \frac{1}{2}O_2 \xrightleftharpoons{(ZrO_2)} H_2O$	1.2	3573	LL
B. Molten Salt Electrolyte				
1. Lithium Iron Sulfide (example)	$2Li + FeS \xrightleftharpoons{(KCl-LiCl)} Fe + Li_2S$	1.6	842	EV, LL
2. Lithium-Tellurium Chloride	$4Li + TeCl_4 \xrightleftharpoons{(KCl-LiCl)} 4LiCl + Te$	3.2	1154	EV
C. Fuel Cells	$H_2 + \frac{1}{2}O_2 \xrightleftharpoons{(4)} H_2O$	1.2	3473	LL

1. Organic solvent containing a dissolved lithium salt
2. β-Al_2O_3 or a Na^+ conductive glass, molten sodium sulfide on the positive side of the solid electrolyte
3. β-Al_2O_3 plus $NaAlCl_4$ on positive side of the solid electrolyte
4. Molten metal carbonate or phosphoric acid
5. Approximate open cell voltage
6. Load-leveling shown as LL, electric vehicle as EV
7. Not rechargeable at this time

upgrading the performance of batteries using these couples. These are discussed below under the individual headings.

Lead-Acid

The theoretical energy density of the lead-acid couple is, with one exception, the lowest of any shown in Table 3. That it is as high as it is presents an interesting sidelight. Its voltage, 2.1 V, is a thermodynamic impossibility using an aqueous electrolyte; the metal negative should corrode actively in the strongly acid electrolyte, releasing hydrogen gas, the positive material should be reduced with the formation of O_2, or (as in this case) both should occur. This does not happen at rates of serious concern because of the high hydrogen and oxygen overvoltages on these electrodes [20].

Increasing the real energy density of the lead-acid battery requires that the weights of all materials in the battery, other than that of the active materials, be reduced to the extent possible. Substitution of plastic battery cases for the hard-rubber ones previously used has significantly improved both weight and volume energy densities. A major contribution to the weight of the battery is found in the lead grids used to sup-support the active materials and to conduct the current with minimum voltage loss. A reduction in the weight of these members is being sought by improved grid design for more effective current collection, by bipolar construction to reduce the weight of internal connectors, and by the use of lead-coated lighter conductive metals such as aluminum and titanium. Reduction of grid corrosion by the use of new lead alloys, the development of improved separators, and the use of tubular construction to prevent shedding of active PbO_2 at the positive electrode represent efforts to improve battery life. The reader is referred to Kirk-Othmer [20] for a detailed description of conventional battery manufacture. Recent developments are described in a number of articles [21, 22, 23]. Based on information found in these references, weight energy densities in the range of 50 Wh/kg at the five-hour rate and a life expectancy in excess of 500 to 700 deep discharge cycles may be achieved in the near future. Gould, Globe Union, and ESB are among the larger companies working on this battery.

Nickel-Zinc

The promise of this system is apparent from its theoretical energy density of 580 Wh/kg. Zinc is the most reactive metal that can be deposited from aqueous solutions and, for this reason, its use as a negative electrode material has been aggressively investigated. As in the case of lead, thermodynamics favors the formation of hydrogen rather than zinc when the electrode is recharged, but the hydrogen overvoltage on zinc is also high enough to permit zinc to be deposited with good coulombic efficiency. Corrosion of zinc in the electrolyte can be greatly reduced by the use of small amounts of mercury and other metals [24, 25].

Two problems have impeded progress in the development of zinc electrodes and, therefore, nickel-zinc batteries: (1) the formation of dendritic zinc deposits on recharge results in shorting through the porous separators, and (2) zinc electrode "shape change" after repeated cycling to deep discharge results in progressive loss of capacity. Both problems appear to be related to the fact that zinc oxide has appreciable solubility in the KOH electrolyte [25, 26]. The problem of dendritic shorting has been attacked by improvement in electrode fabrication techniques, new separators, and by additives which reduce the solubility of $Zn(OH)_2$ in the electrolyte [21, 27, 28]. The problem of shape change, a progressive thickening of the zinc deposit in the center and lower portion of the electrode, is probably the major factor today in limiting the useful electrode life. The effective capacity decreases with cycling, and the life of the electrode must be defined by the percentage of initial capacity loss which can be tolerated. The situation is further complicated by the fact that the rate of capacity loss increases with depth of discharge and discharge rate. It follows that electrode life can only be defined by an arbitrary set of circumstances. With this caveat, the life of nickel-zinc batteries for vehicular use is currently estimated at about 300 to 500 cycles [13, 27].

A variety of unconventional means have been proposed to avoid the problems of zinc electrode life which, in one form or another, use some form of internal agitation of the electrolyte or action on the deposit itself [29, 30, 31]. None of these methods is known to be under active commercial development at this time.

While the problems of the zinc electrode have been discussed above in terms of life, the problems of its companion electrode in the nickel-zinc couple are more pragmatic. These electrodes have demonstrated a useful life of thousands of cycles in nickel-cadmium and nickel-iron batteries. High rate nickel oxide electrodes use sintered nickel plaques to contain the active material and to serve as the current collector. At least as much nickel is present in this structure as in the active material.

Efforts are being made to substitute a mixture of graphite powder and organic binder for this relatively expensive structure [27, 32]. If this approach is successful, this would both reduce the cost and increase the energy density of this battery. This development has been cited as the basis for the development of vehicle batteries of this type with energy densities as high as 70 to 90 Wh/kg and lives of 500 or more cycles [32]. Eagle-Picher has reported the achievement of 40 to 50 Wh/kg and has projected costs of $50 to $100/kWh [34]. A number of other companies, including Gould and Yardney, are known to be working on this system.

Nickel-Iron

This battery is one of the oldest secondary batteries still in use. Called the Edison cell after its inventor, it offers exceptional life and durability and is one of the few systems shown for which 2,000 deep discharge cycles can virtually be guaranteed [21, 33]. Unfortunately, its other properties have been less desirable. In their earlier forms, these batteries were characterized by poor charge acceptance at the negative electrode with much hydrogen formation, excessive rates of self-discharge, and poor high rate discharge characteristics, particularly at low temperatures [20]. The high gassing rate could present a serious safety hazard, particularly for large load-leveling battery plants. Westinghouse Electric is known to have been working on this battery, but the writer is not aware of any recent publications by that company which indicate present commercial progress. Gross, however, cites a private communication indicating achievement of an energy density of 48 Wh/kg, 900 cycles to complete discharge, and an estimated cost of $100/kWh [21].

Nickel-Hydrogen

This couple is included here to illustrate the problem of using a stored gaseous reactant. The theoretical energy density is high but the hydrogen is physically stored in a pressure vessel which also houses the electrode assembly [21, 35, 36]. The weight and cost of the containment vessel would appear to limit consideration of this system to load-leveling use although Miller describes it as a candidate for vehicle use [34]. An equally serious consideration for either application, however, is the projected cost of $100 to $200/kWh. The development of low-cost nickel oxide electrodes of the type described above for the nickel-zinc battery, as well as low cost catalysts for the H_2 electrode, may determine the future prospects for this battery. The possibility of storing hydrogen in a condensed form as a hydride of an iron-titanium alloy has also been proposed [34]. One study has indicated that the cost of this expedient is roughly the same as that using a pressure vessel [9].

Lithium Negative Batteries

Because of its low molecular weight and the high voltages obtained with many couples using this metal, lithium presents both a unique opportunity and a challenge. Thermodynamically unstable in contact with water, it nevertheless is being used in this environment by Lockheed in conjunction with water and an inert positive electrode [37]. The reaction product, $Li(OH)_2$, has a limited solubility, and the solid barrier layer formed on the surface of the electrode permits discharge of solid lithium electrodes with rather good electrochemical utilization. However, these electrodes cannot be recharged electrochemically from an aqueous electrolyte.

The more conventional approach to the use of lithium electrodes lies in the use of electrolytes which are thermodynamically or practically stable to this metal. At ambient temperatures, this involves the use of aprotic solvents such as propylene carbonate containing dissolved lithium salts. Active positive materials have included a wide variety of metal oxides, sulfides, and fluorides [38]. The low conductivity of these electrolytes has, however, limited the use of these cells to low power applications such as electronic equipment. Discharge at higher rates has been achieved by the use of thionyl chloride as the electrolyte [39]. None of these battery systems have been found to be usefully reversible, but Exxon has reported that the lithium-aluminum/TiS_2/organic electrolyte system is both reversible and capable of high enough power output to be useful for electric vehicles [40]. The TiS_2 positive active material represents a new class of active positive materials for which discharge does not involve reduction of the original atomic members of the compound in the positive electrode. Lithium ions enter the positive material via intercalation to form the final product $LiTiS_2$. Costs of $34/kWh and an energy density of 130 Wh/kg have been projected for batteries of this type.

Zinc-Halogen

The zinc halogen systems are particularly interesting because they provide a means of storing gaseous reactants in a reduced volume as a solid or a liquid. The zinc-chlorine system under exploration by Energy

Development Associates (EDA), a joint venture of Gulf and Western Industries and Occidental Petroleum, stores chlorine gas as a solid hydrate, $Cl_2(H_2O)_6$, stable at a temperature of ~10°C [41]. The zinc-bromine system uses bromine-organic amine complexes to store the product as a liquid, insoluble in the electrolyte, $ZnBr_2$ [42]. This battery concept is being developed by Eco-Control, Inc. [42].

Zinc metal can be deposited on an inert electrode from concentrated solutions of zinc chloride or bromide without encountering the dendritic deposition of zinc from alkaline solutions. The chlorine gas generated at an inert electrode during charging is precipitated as solid particles of $Cl_2(H_2O)_6$ in the refrigerated electrolyte and is stored in an external container. Bromine reacts with a complex organic amine dissolved in the electrolyte and is also stored externally. Chlorine gas can be liberated from the solid hydrate for discharge by warming the electrolyte. The liquid bromine complex can be pumped through the cells from the storage container.

Energy densities as high as 130 Wh/kg and costs as low as $15/kWh have been projected for the zinc-chlorine battery. If these projections are met in fully developed batteries, this system would find ready acceptance for both vehicle and utility use. The zinc-bromine system is in an early stage of development and no energy density or cost projections are known to have been made for fully developed batteries of this type. Potential development problems include the need for pumps and heat exchangers which could reduce overall energy efficiency through parasitic power losses, possible release of chlorine gas if power for the refrigerated chlorine hydrate storage unit were lost, and self-discharge of the zinc metal deposit in contact with halogen gases dissolved in the electrolyte.

Hydrogen-Oxygen (Air) Fuel Cells

Although considerable work was done on ambient temperature fuel cells in the past, none is known to have been carried out in recent years. Part of the problem here arises from the fact that the oxygen electrode reaction is highly irreversible with the formation of the perhydroxyl ion, O_2H^-, being favored unless expensive catalytic materials such as platinum or silver are used which accelerate decomposition of this ion.

Equally important are the problems encountered in supplying oxygen in the form of an ordinary atmosphere. Air contains CO_2 which reacts with the alkaline electrolyte, forming K_2CO_3 which reduces the ionic conductivity of the electrolyte and plugs the pores of hydrophobic (Teflon) membranes used to provide an interface between the gas (air) and fluid (aqueous KOH) phases. This reduces the oxygen electrode efficiency and causes "weeping" of electrolyte through the hydrophobic porous barrier. These problems are common to all metal-air cells such as the zinc-air battery discussed below.

Zinc-Air

A new approach to the problems of both the zinc and air electrodes has been sought in the work on a zinc-air battery now under development by Marcoussis of France. The negative electrode consists of a slurry of zinc particles suspended in the flowing electrolyte. These particles are in fortuitous contact with an inert metal current conductor. The air electrode uses inexpensive, high surface area carbon as the catalytic surface at the expense of increased electrode polarization. The incoming air is purged of CO_2 by passage through the alkaline electrolyte before it contacts the air electrode. Zinc powder is regenerated electrochemically from the alkaline electrolyte, and the inexpensive air electrode is replaced after its efficiency has declined with use. Projected energy densities of 85 to 100 Wh/kg have been made. The operating costs using this battery, including amortization, have been described as less than can be achieved with lead-acid batteries and comparable to those presently obtained with gasoline engines [43].

Redox Cells

Redox cells are being considered for utility energy storage systems. Oxidizable and reducible salts in aqueous solutions are pumped past inert positive and negative electrodes during discharge to produce electrical energy. The spent solutions are returned to storage tanks. The reverse process occurs during charge. The low theoretical energy density shown for the example in Table 3 illustrates the fact that this approach is suitable only for load-leveling use. The incentive here lies in the potentially low cost of these systems and the absence of problems with deposition of metallic deposits at the negative electrodes (classically, the oxidized and reduced species are soluble at both electrodes). Thaller has proposed the use of the $Fe^{2+}/Fe^{3+}//Ti^{3+}/Ti^{4+}$ couple [44]. Ion selective membranes would be required to prevent mixing of different ionic species in

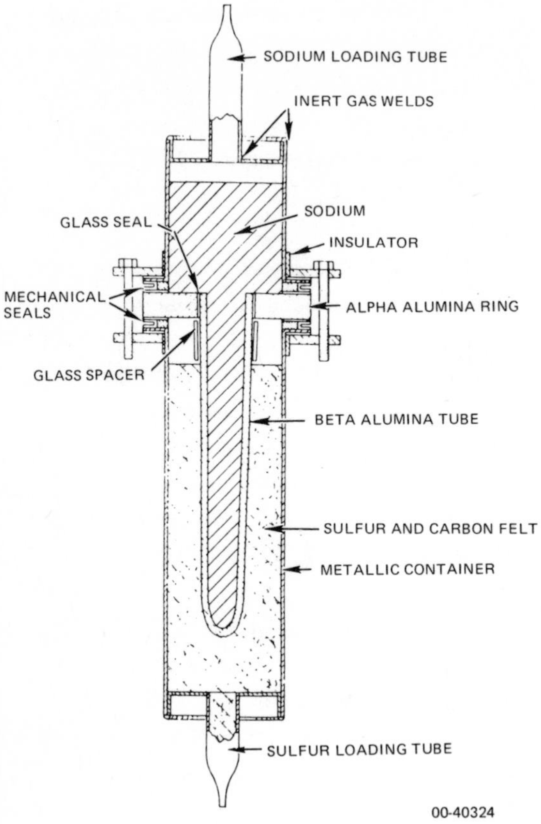

Figure 5. Prototype Sodium-Sulfur Cell of 16 Ah Capacity

certain impurities, such as silicates or potassium ions, is necessary for extended satisfactory performance of this cell element [52]. The highly reactive materials must be protected from contact with oxygen and moisture in the atmosphere, and hermetic, electrically insulating seals must therefore be provided between the cell case and the ceramic components. Corrosion of the outer metal case has been reported as a major problem. An electronically conductive ceramic case has been considered as a possible solution to this difficulty [48, 52].

Whereas Ford appears to have restricted cell design to the single tube design shown on Figure 5, General Electric has proposed a multiplicity of tubes in a single header. Cell sizes have ranged from 30 to 50 Wh at these companies. The life of the cells are defined as Ah passed per cm^2

of tube area. General Electric has reported a life on this basis in excess of 300 Ah/cm^2 [48]. Ford has exceeded 1000 Ah/cm^2 with a small cell with no metal parts [52]. Both Ford and General Electric are adhering to cost goals of ~\$20/kWh for load leveling use. Energy densities up to 150 Wh/kg (around three times that estimated for advanced lead-acid batteries) are projected for fully engineered vehicle batteries [51].

Lithium - Iron Sulfide

This type of battery actually represents a family of cell couples depending on the choice of the active materials used in the negative and positive electrodes. The choices made to date differ with the principal investigators, Atomics International Division of Rockwell International, Argonne National Laboratory, and General Motors. Three possibilities exist with respect to the negative electrode. Liquid elemental lithium offers the highest possible cell voltage, and therefore has been investigated at all three of these facilities. However, failure to achieve reliable containment of the liquid metal, self-discharge of cells using the elemental material, and problems with attack on ceramic separators, has reduced interest in this approach [53]. The use of lithium alloys, solid at the cell operating temperature of 400 to 450°C, reduces the severity of these problems, but a loss in cell voltage results. Argonne National Laboratory has concentrated its efforts on the use of the lithium aluminum alloy, LiAl [54]. The use of this alloy reduces the cell voltage about 300 mV from that obtainable with liquid lithium. The use of solid lithium silicon alloys for this purpose was first reported by Atomics International [55], but General Motors is also investigating this material [56]. A number of compositions ranging from Li_5Si to Li_2Si have been identified in the discharge of this class of compound. The most lithium-rich of these, Li_5Si, discharges in four steps ranging from ~+50 mV to 340 mV positive to lithium. Somewhat higher theoretical capacities, in terms of Ah/gm of alloy and a lower average voltage loss vs lithium, can therefore be achieved with lithium-silicon alloys than with the lithium-aluminum alloy.

Two iron sulfides may be used. FeS discharges at a single voltage plateau, while FeS_2 discharges at two voltage levels, approximately the first half of its capacity being discharged at 2.05 V (vs lithium) and the second half at 1.65 V (vs lithium). While higher energy densities are achievable with FeS_2, corrosion problems with metal components of the positive electrode are more severe at the higher potentials required to charge this positive electrode. For vehicles where high energy density is required, FeS_2 offers a definite advantage. For load leveling, where cost is most important, FeS may be preferred because less expensive metals can be used in making the electrode structures [55].

The theoretical energy densities for this system depend on the choice of active materials used in the negative and positive electrodes. These range from a low value of 693 Wh/kg, where LiAl electrodes are combined with FeS, to a high of 1038 Wh/kg using an Li_5Si/FeS_2 couple. Emphasis is being placed on vehicle batteries by Argonne National Laboratory and General Motors. Energy density projections for fully developed vehicle batteries range from 200 Wh/kg [32] to more modest near-term goals of 120 Wh/kg set by Argonne National Laboratory [54]. Emphasis has been placed on the development of load-leveling batteries at Atomics International with cost goals of $25 to $30/kWh and a functional life of 2000 cycles.

High Temperature Fuel Cells

Fuel cells differ in one major respect from the battery systems discussed above in that they receive flowing gaseous reactants containing impurities which can have an adverse effect on the cell operation. Of these impurities, CO_2 is certainly one of the more important, and the choice of the electrolyte is dictated by the need for compatibility with carbon dioxide in both reactant gases. Alkaline (KOH) electrolytes will absorb CO_2 from air or from hydrogen derived by cracking hydrocarbon feed stocks. The use of phosphoric acid as the electrolyte avoids this problem but the oxygen electrode kinetics are less favorable in acid electrolytes [10]. An alternative approach, the use of a molten carbonate electrolyte immobilized in a porous ceramic plate or tube is also under active investigation [57, 58]. It is important to note that this "solid" electrolyte is solid only in a physical sense and bears no resemblance in function to the solid β-alumina electrolyte used in the sodium-sulfur cell. A true ion-conducting solid electrolyte was investigated by the Westinghouse Electric Corporation a number of years ago. A "doped" (CaO, Y_2O_3) ZrO_2 electrolyte was used in this work [59]. Although all three types of cells involve the hydrogen-oxygen (air) couple, the electrode reactions differ markedly and are, therefore, interesting in their diversity. These are shown below for each of the types described above.

Phosphoric Acid Electrolyte

$$H_2 \rightleftharpoons 2H^+ + 2e$$
$$1/2\ O_2 + H_2O + 2e \rightleftharpoons 2OH^-$$
$$\overline{1/2\ O_2 + H_2 \rightleftharpoons H_2}$$

Molten Carbonate

$$H_2 + CO_3^{=} \rightleftharpoons H_2O + CO_2 + 2e^-$$
$$1/2\ O_2 + CO_2 + 2e \rightleftharpoons CO_3^{=}$$
$$\overline{1/2\ O_2 + H_2 \rightleftharpoons H_2O}$$

Solid Electrolyte

$$H_2 + O^{=} \rightleftharpoons H_2O + 2e$$
$$1/2\ O_2 + 2e \rightleftharpoons O^{=}$$
$$\overline{1/2\ O_2 + H_2 \rightleftharpoons H_2O}$$

Thus, while the overall reactions are the same, both the reactants and/or the conductive ions are different in each case. The phosphoric acid cells operate at 150 °C, the molten carbonate cells at about 500 °C, while the oxide ceramic cells require temperatures as high as 1000 °C for adequate $O^{=}$ conductivity. At this time, a large scale program at United Technology to develop utility power systems using phosphoric acid and molten carbonate fuel and conductive oxide cells are being funded by the Electric Power Research Institute [10]. Molten carbonate systems are being investigated by the Institute of Gas Technology [58].

SUMMARY

Anticipated changes in energy source patterns have created new incentives for the development of advanced electrochemical energy systems. At the same time, these new systems must meet difficult performance and cost goals. For these reasons, new efforts are being made to overcome complex fundamental problems in electrode kinetics, corrosion, materials compatibility, etc, which have hindered development of these systems up to this time.

It has been shown that the rewards for successful performance are large. So, indeed, are the costs. The problem here is to compress development time into an exceedingly short time frame. To place this matter into some historical perspective, it should be observed that the lead-acid battery originated with Plante in 1859. Today, over 100 years later, improvements are still being sought.

A recent article notes that Energy Development Associates (EDA) has expended more than nine years and $10 million to develop the zinc-chlorine battery [60]. Electric Power Research Institute and EDA are preparing to spend jointly over $7.6 million over the next 39 months with the goal of reaching a 10 MWh zinc-chlorine load-leveling battery system. At present, a 20 kWh system of this type is easily the largest advanced battery system known to be in operation. Large sums are also being expended by the Energy Research and Development Administration for a variety of systems, including the lithium - iron sulfide batteries. ERDA and EPRI are also funding the larger portion of a $42 million contract with United Technologies for the development of a 4.8 MW fuel cell system [61].

The Electric and Hybrid Vehicle Research and Development Act passed last year by Congress specifies the purchase of up to 2,500 improved electric vehicles by July 1978, and up to 5,000 vehicles with advanced batteries and vehicle designs by April 1981. This development can be expected to accelerate activities on the part of private corporations who are also spending large unreported amounts of research and development money on battery systems in which they have a proprietary interest. The Manhattan Project for the development of the atom bomb succeeded in "compressing history" by the expenditure of large sums of money. It remains to be seen to what degree the same approach can speed the development of advanced electrochemical systems for a more beneficient purpose.

REFERENCES

1. Minerals Yearbook, Volume 1, Bureau of Mines, U.S. Gov. Printing Office, Wash. D.C., 1974

2. J. Cook, "The Invisible Crisis," Forbes Magazine, July 15, 1976, p. 26

3. Gas Supply Review, Supplement, American Gas Association, Gas Supply Committee, July 15, 1975

4. United States Mineral Resources, Geological Survey, Professional Paper 820, ed. D.A. Brobst and W.P. Pratt, U.S. Gov. Printing Office, Wash. D.C., 1973

5. "Uranium Resources to Meet Long-Term Uranium Requirements," Special Report EPRI SR-15, November 1974

6. "Advanced Nuclear Reactors," Report ERDA-46, U.S. Gov. Printing Office, Wash. D.C., September 1975

7. J.R. Birk and F.R. Kalhammer, "Secondary Batteries for Load-Leveling," Proc. of Symp. and Workshop on Advanced Battery Research and Design, Argonne Nat'l Lab., Argonne, IL, March 1976, p. A-2

8. N.P. Yao and J.R. Birk, "Battery Energy Storage for Utility Load-Leveling and Electric Vehicles: A Review of Advanced Secondary Batteries," 10th Intersociety Energy Conversion Engineering Conf., Newark, DE, August 1975, p. 1107

9. "An Assessment of Energy Storage Systems Suitable for Use by Electric Utilities," Report EM-264, EPRI, Volume II, July 1976

10. A.P. Fickett, "Fuel Cells: Versatile Power Generators," EPRI Journal, April 1976, p. 14

11. A.J. Konopka and D.P. Gregory, "Hydrogen Production by Electrolysis: Present and Future," 10th Intersociety Energy Engineering Conversion Conf., Newark, DE, August 1975, p. 1184

12. E.C. Jerabek, "Load-Leveling with Electric Vehicles in the Urban Environment," 11th Intersociety Energy Conversion and Engineering Conf., State Line, NV, September 1976, p. 382

13. A.R. Landgrebe, "Secondary Batteries for Electric Vehicles," Proceedings of Symposium and Workshop on Advanced Battery Research and Design, Argonne Nat'l Laboratory, Argonne, IL, March 1976, p. A-19

14. H.J. Schwartz, "The Requirements for Batteries for Electric Vehicles," 27th Power Sources Symp., June 1976, p. 20

15. A.R. Landgrebe, K. Klunder, and N.P. Yao, "Federal Battery Program for Transportation Uses," 27th Power Sources Symp., June, 1976, p. 23

16. M. C. Yew and D. E. McCulloch, "Small Electric Vehicle Considerations in View of Performance and Energy Usage," 11th Intersociety Energy Conversion Engineering Conf., State Line, NV, September 1976, p. 363

17. J. J. Reilly, K. C. Hoffman, G. Strickland and R. H. Wiswall, "Iron Titanium Hydride as a Source of Hydrogen Fuel for Stationary and Automotive Applications," 26th Power Sources Symp., April and May 1974, p. 11

18. A. W. Liles and G. P. Fetterman, "Selection of Driving Cycles for Electric Vehicles of the 1990's," 11th Intersociety Energy Conversion Engineering Conf., State Line, NV, September 1976, p. 390

19. "The Automobile and Air Pollution: A Program for Progress," Part II, U. S. Dept. of Commerce, U. S. Gov. Printing Office, Wash. D. C., December 1967, p. 63

20. Kirk-Othmer Encyclopedia of Chemical Technology, Volume 3, John Wiley and Sons, Inc, NY, 1964, p. 161

21. S. Gross, "Review of Candidate Batteries for Electric Vehicles," Energy Conversion, 15, 1976, p. 95

22. N. J. Maskalick, J. T. Brown and G. A. Monito, "The Case for Lead-Acid Storage Battery Peaking Systems," 10th Intersociety Energy Conversion and Engineering Conf., Newark, DE, August 1975, p. 1135

23. D. W. Kassekert, A. O. Isenberg and J. T. Brown, "High Energy Density Bipolar Lead-Acid Battery for Electric Vehicle Propulsion," 11th Intersociety Energy Conversion Engineering Conf., State Line, NV, September 1976, p. 411

24. T. P. Dirkse and R. Timmer, "The Corrosion of Zinc in KOH Solutions," J. Electrochem Soc. 116, 1969, p. 162

25. J. E. Oxley, C. W. Fleischmann and H. G. Oswin, "Improved Zinc Electrodes for Secondary Batteries," Proc. of 20th Annual Power Sources Conf., Atlantic City, NJ, 1966, p. 123

26. J. W. Diggle, A. R. Despic and J. O'M. Bockris, "The Mechanism of the Dendritic Electrocrystallization of Zinc," J. Electrochem. Soc. 116, 1969, p. 1503

27. A. Charkey, "Advances in Component Technology for Nickel-Zinc Cells," 11th Intersociety Energy Conversion Engineering Conf., State Line, NV, September 1976, P. 452

28. A. Fleischer and J. Lander, "Zinc-Silver-Oxide Batteries," John Wiley, NY, 1971

29. A. von Krusenstierna, "Method and Apparatus for Avoiding Dendrite Formation when Charging Accumulator Batteries," U. S. Patent 3,923,550, December 2, 1975

30. Z. Stachurski, "Rechargeable Current-Generating Electrochemical System with Wiper Means," U. S. Patent 3,440,098, April 22, 1969

31. L. R. McCoy, "Secondary Battery with Movable Shutter Means Between Fixed Electrodes," U. S. Patent 3,762,959, October 2, 1973

32. E. J. Cairns and J. McBreen, "Batteries Power Urban Autos," Industrial Research, June 1975, p. 57

33. S. V. Falk and A. J. Salkind, "Alkaline Storage Batteries," John Wiley and Sons, Inc., NY, 1969, p. 319

34. L. E. Miller and R. A. Brown, "Nickel Battery Systems for Electric Vehicles," Proc. of 27th Power Sources Symp., June 1976, p. 16

35. M. Klein, "Nickel-Hydrogen Secondary Battery," 10th Intersociety Energy Conversion Engineering Conf., Newark, DE, August 1976, p. 803

36. L. E. Miller, "Nickel-Hydrogen as an Alternative to Lead-Acid and Nickel-Cadmium Systems in Nonspace Applications," 10th Intersociety Energy Conversion Engineering Conf., August 1976, p. 807

37. H. J. Halberstadt, E. L. Littauer and E. S. Schaller, "Physical and Economic Characteristics of the Lithium-Water Marine Battery," 10th Intersociety Energy Conversion Engineering Conf., Newark, DE, August 1975, p. 1120

38. S. Gilman, "An Overview of the Primary Lithium Battery Program," Proc. of 26th Power Sources Symp., 1974, p. 28

39. N. Marincic, A. Lombardi and C. R. Schlaikjer, "Progress in the Development of Lithium Inorganic Batteries," Proc. of 27th Power Sources Symp., June 1976, p. 37

40. L. H. Gaines, R. W. Francis, G. H. Newman and B. M. L. Rao, "Ambient Temperature Electric Vehicle Batteries Based on Lithium and Titanium Disulfide," 11th Intersociety Energy Conversion Engineering Conf., State Line, NV, September 1976, p. 418

41. P. C. Symons, "Batteries for Practical Electric Cars," paper presented at International Automotive Engineering Congress, Detroit, MI, January 1973

42. M. Walsh, F. Walsh and D. Crouse, "Zinc-Polybromide Batteries," 10th Intersociety Energy Conversion Engineering Conf., Newark, DE, August 1976, p. 1141

43. A. J. Appleby, J. P. Pompon and J. Jacquier, "Economic and Technical Aspects of the C. G. E. Zn-Air Vehicle Battery," 10th Intersociety Energy Conversion Engineering Conf., Newark, DE, August 1975, p. 811

44. L. Thaller, "Electrochemically Rechargeable Redox Flow Cells," 9th Intersociety Energy Conversion Engineering Conf., San Francisco, California, August 1974, page 924

45. Batelle-Geneva Research Center, "Electrochemical Energy Storage for Load-Leveling Applications," proposal issued by Center, 1975

46. "Sodium Chloride Battery Development Program for Load-Leveling," Report EPRI-230, prepared for EPRI by ESB, Inc., January to December 1975

47. J. C. Schaefer, T. M. Noveske, J. S. Thompson and B. Profeta, "The ESB-Sohio Carb-Tek Molten Salt Cell," 10th Intersociety Energy Conversion Engineering Conf., Newark, DE, August 1975, p. 649

48. "Development of Sodium-Sulfur Batteries for Utility Application," Report EM-26, prepared for EPRI by the General Electric Co., December 1976

49. C. A. Levine, "Progress in the Development of the Hollow Fiber Sodium-Sulfur Secondary Cell," 10th Intersociety Energy Conversion Engineering Conf., Newark, DE, August 1976, p. 621

50. R, M. Dell, J. L. Sudworth and I. W. Jones, "Sodium/Sulfur Battery Development in the United Kingdom, 11th Intersociety Energy Conversion Engineering Conf., State Line, NV, September 1976, p. 503

51. J. L. Sudworth, "Sodium/Sulfur Batteries for Rail Traction," 10th Intersociety Energy Conversion Conf., Newark, DE, August 1976, p. 616

52. "Research on Electrodes and Electrolyte for the Ford Sodium-Sulfur Battery," Semi-Annual Report for June 1975 to December 1975, prepared by Ford Motor Co. for RANN Division of NSF, Wash. D. C.

53. L. R. McCoy and L. A. Heredy, "Development Status of Lithium-Silicon-Iron Sulfide Load Leveling Batteries," 11th Intersociety Energy Conversion Engineering Conf., State Line, NV, September 1976, p. 485

54. E. T. Gay, T. D. Kaun, and F. J. Martino, "Review of Electrode Designs and Fabrication Techniques for Lithium-Aluminum/Iron Sulfide Cells," 11th Intersociety Energy Conversion Engineering Conf., State Line, NV, September 1976, p. 477

55. S. Sudar, L. R. McCoy, and L. A. Heredy, "Rechargeable Lithium/Iron Sulfide Battery," 10th Intersociety Energy Conversion Engineering Conf., Newark, DE, August 1975, p. 642

56. R. A. Sharma and R. N. Seeforth, "Thermodynamic Properties of the Lithium-Silicon System," J. Electrochem. Soc. 123, 1976, p. 1763

57. I. Trachtenberg, "Electrode Processes in Molten Carbonate Fuel Cells," Advances in Chemistry Series, Number 47, Am. Chem. Soc., 1965, p. 232

58. B. S. Baker, L. G. Marianowski, J. Meek, and H. R. Linden, "High Temperature Natural Gas Fuel Cells," Advanced in Chemistry Series, Number 47, Am. Chem. Soc., 1965, p. 247

59. D.H. Archer, J.J. Alles, W.A. English, L. Elikan, E.F. Sverdrup, and R.L. Zahradnik, "Westinghouse Solid-Electrolyte Fuel Cell," Advances in Chemistry Series, Number 47, Am. Chem. Soc., p. 332

60. Advanced Battery Technology, Publ. Robert Morey Assoc., Dana Point, CA, March 1977

61. Ibid, October 1976

We shall now discuss the various photovoltaic effects in electrochemical cells, from the viewpoint of where the primary excitation occurs.

Photoeffects Caused by Excited Molecules

We have already mentioned that an excited molecule has a higher reduction power as well as a higher oxidation power. In electrochemical language this means that an excited molecule cannot be characterized by a single redox potential, since one has to distinguish between the redox potential for the reducing action and the redox potential for the oxidizing action of the excited molecule [4]. Electron energies in excited states are best described by the spectroscopic term scheme used in Fig. 1. The electrochemical scale of redox potentials represents the free energy of electrons in a system and is equivalent to the Fermi energy of electrons in a solid. It is closely related to such a spectroscopic term scheme and can be transformed into such a scale by a simple operation. The electrochemical potential scale relates the free energy of the electrons in a particular redox reaction to the free energy of the electrons in a particular redox reaction to the free energy of electrons of a reference redox reaction, usually the standard hydrogen electrode. In spectroscopy and in solid-state theory, the energy levels of the electrons are often related to the vacuum level of the electron as the reference state. The ionization energy gives the binding energy in the ground state; the binding energies in the excited states are obtained by subtracting the excitation energies. One can define a redox potential scale for electrons in the very same way by using the difference between the free energy of electrons in a redox system and electrons at the vacuum level as a scale parameter. For this transformation one has only to know the binding free energy of electrons in the standard hydrogen electrode, which has been calculated to be about -4.5 eV [5] or -4.3 [6]. This is the same as the position of the Fermi energy in the metal which acts as electron donor and acceptor in the standard hydrogen electrode. Unfortunately, the sign convention in the electrochemical redox potential scale is just opposite to that of the absolute scale which makes the picture somewhat confusing. However, it is only a simple linear transformation between the electrochemical scale, E_{redox}, and the absolute scale of redox potentials, $E_{F,redox}$, as given in the following formula:

$$E_{F,redox} = \text{const} - E_{redox}; \quad \text{const} = -4.5 \text{ eV}$$

A comparison between the absolute scale and the electrochemical scale is shown in Fig. 2.

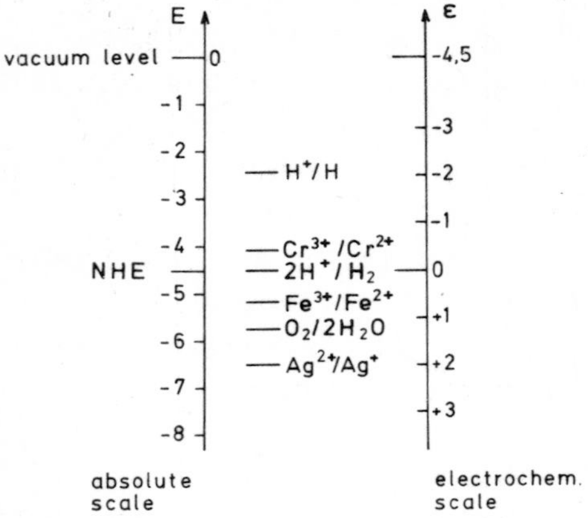

Fig. 2. Correlation between absolute and electrochemical scale of redox potentials.

We can now discuss more quantitatively the relation between the redox potential of an excited molecule and a molecule in the ground state. The energy term scheme of Fig. 3 shows the redox levels for a molecule which can either be oxidized or reduced. The energy levels correspond to an electron transfer to or from the vacuum level under Franck-Condon conditions, that is, ionization energies or electron affinities without structural reorganisation. For oxidation as shown on the left side of the figure an electron has to be removed from the highest occupied orbital. This orbital therefore represents the donor orbital. The result is a vacant state in this orbital which can now act as an acceptor state. Due to the different interaction with the surroundings, the energy that will be obtained if an electron is brought from the vacuum level to this vacant level will be somewhat smaller than the energy necessary to remove the electron

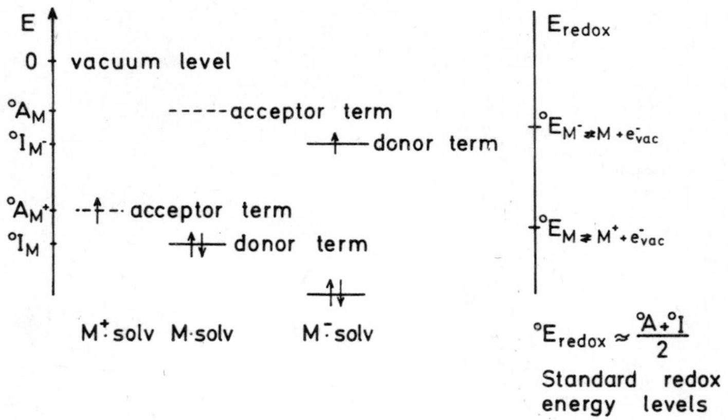

Fig. 3. Ionization energies (°I) and electron affinities (°A) of a molecule M in solution in its ground state for oxidation and reduction at Franck-Condon conditions. The corresponding redox potentials in the absolute scale are shown at the right side.

from the occupied level. This is indicated in the figure by a small upwards shift of this acceptor term. The energy level which represents the average free energy of electrons in this redox reaction, that is, the redox potential or the Fermi energy of this system, is midway between these two energy terms, as is indicated on the right side of this figure. A more detailed description of energy term distributions and their relation to redox potentials is given elsewhere [7].

If this molecule has to be reduced, the lowest unoccupied orbital has to act as the acceptor term. The energy of this term is far above the occupied energy levels of the molecule M as seen in the middle of the figure. In the reduced state this acceptor term contains an electron and acts as the donor level for the redox system M/M^-. Again, due to the interaction in the surroundings, the occupied level is somewhat below the unoccupied level, and the redox potential for the reduction of this molecule is just between these two terms as shown on the right side of the figure.

In some cases both redox potentials of a molecule can be obtained experimentally, whereas in many cases only one term can be measured and the other is not accessible, due to irreversible reactions connected with such an electron transfer reaction. In the latter systems rough approximations of the unknown redox potential may be obtained from theoretical considerations. If these two redox potentials in the ground state are known we can easily derive the position of the redox potential for the excited molecule using the procedure shown in Fig. 4.

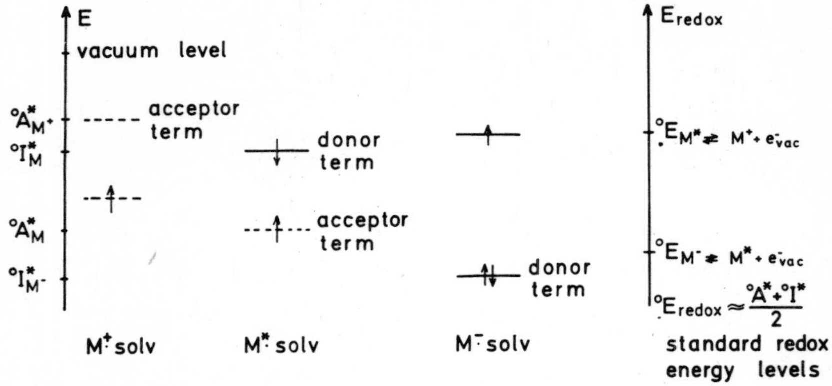

Fig. 4. The energy diagram of Fig. 3 for the molecule M in the excited state M*.

In the middle of Fig. 4 the term scheme of the neutral molecule is shown in its excited state. One sees that in this case, compared to the ground state, the donor term for the oxidation of this molecule is shifted upwards by the excitation energy. This is also the case for the acceptor term of the oxidized molecule, if we assume that the excited molecule is formed by this electron transfer. Therefore, the donor, as well as the acceptor, states of the excited molecule are shifted upwards by the excitation energy in the direction of oxidation. As a consequence the redox potential for the oxidation of the molecule in its excited state is also shifted upwards by the excitation energy. From the redox potential for the oxidation of the molecule in the ground state one obtains in this way the redox potential for oxidation in the excited state.

For reduction we have just the opposite situation. The acceptor term in the excited molecule, as well as the donor term of the reduced molecule, is shifted downwards if the excited state has to be formed by removal of an electron. This means that the redox potential of the excited molecule is shifted downwards (in the direction of reduction) by the excitation energy, relative to the redox potential of the molecule in its ground state, on the absolute scale of redox potentials. Both redox potentials of the excited molecules are shown on the right side of Fig. 4.

It should be mentioned that this picture has been somewhat simplified to make it easily understandable. The shifts of the energy levels for

oxidation and reduction due to the different internal electronic interactions in the molecule are neglected. Also, the excitation energy that causes the shift in the redox energy levels is not the energy that is directly absorbed by the molecule, but rather it is the energy that remains stored in the excited molecule after internal relaxation. However, the simplified picture is sufficient for understanding the principle. A more precise discussion is given in [8].

Photocurrents Caused by Photo-Redox Reactions in Homogeneous Solution

We have seen that an excited molecule has two different redox potentials. Whether it is oxidized or reduced in the excited state depends on the reaction partner it finds in its surroundings. We can anticipate two types of reactions which are shown in the reaction scheme of Fig. 5, where two redox couples with different standard redox potentials are supposed to be at equilibrium in the dark. By illumination, one component of one of the redox couples may absorb the light and change its redox potential as discussed above. As a result the redox reaction will proceed now in one or the other direction.

photochemical redox reactions

$$Red_1 + h\nu \longrightarrow Red_1^*$$
$$Red_1^* + Ox_2 \xrightarrow{fast} Ox_1 + Red_2$$
$$Ox_1 + Red_2 \xrightarrow{slow} Red_1 + Ox_2$$

consequence: $E_{2,0} > E_{1,0}$ at illumination

$$\text{or } Ox_1 + h\nu \longrightarrow Ox_1^*$$
$$Ox_1^* + Red_2 \xrightarrow{fast} Red_1 + Ox_2$$
$$Red_1 + Ox_2 \xrightarrow{slow} Ox_1 + Red_2$$

consequence: $E_{2,0} < E_{1,0}$ at illumination

Fig. 5. Photochemical redox reactions.

The two redox systems will reach a steady state where, due to the different chemical composition of the solution, the redox potentials will differ. Some time will be needed until the steady state has been reached. This is schematically shown in Fig. 6. How far the two redox potentials will differ depends on the illumination intensity and on the rate of the

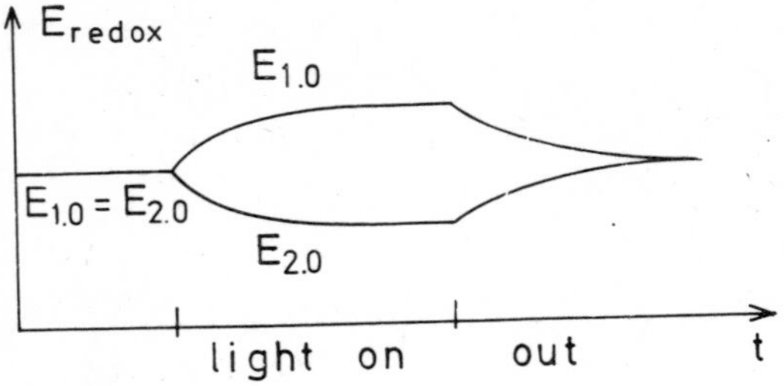

Fig. 6. Transient of the redox potentials of two redox couples if the oxidized component Ox_1^* reacts in the excited state with the reduced component Red_2.

reverse reaction. The difference will be greater the slower the reverse reactions are, and the time required to reach the steady state will also increase. Consequently, large effects can be achieved only if the reverse reactions are sufficiently slow.

Such a photo-redox reaction where a large deviation from equilibrium can be achieved has been described by Rabinowitch [9], namely, the oxidation of Fe^{2+}-ions by excited thionine molecules. Rabinowitch observed that a photocurrent flows between an illuminated and a dark electrode immersed into a solution of thionine and Fe^{2+}-ions. He has called this the "photogalvanic effect". Such systems have recently been studied in much detail by Lichtin [10]. The mechanism of this reaction (somewhat simplified) is given in Fig. 7. Lichtin's group has tried to exploit this effect for solar energy conversion. A solar cell using a transparent electrode in order to reach the highest photoeffects close to the electrode surface is shown in Fig. 8 [11].

The difference in redox potentials in such a homogeneous solution only can be exploited for the generation of photocurrents if the two electrodes immersed into the solution act selectively for one of the redox reactions. Otherwise the electrodes would only act as a catalyst for the reverse reaction and no, or a very small, external current would flow. This can be understood if one compares the partial current voltage curves for the two redox reactions at such an electrode in the dark and under illumination [12] as is shown in Fig. 9. Both current voltage curves for the two different

Photooxidation of Fe^{2+} by Thionine (after Lichtin)

$$Th \xrightarrow{h\nu} Th^*$$

$$Th^* + Fe^{2+} + H^+ \longrightarrow Th{\cdot}H + Fe^{3+}$$

$$2\,Th{\cdot}H \xrightarrow{fast} Th + ThH_2$$

$$Th{\cdot}H + Fe^{3+} \xrightarrow{fast} Th + Fe^{2+} + H^+$$

$$ThH_2 + Fe^{3+} \xrightarrow{slow} Th{\cdot}H + Fe^{2+} + H^+$$

$$ThH_2 + Th \xrightarrow{slow} 2\,Th{\cdot}H$$

Th =

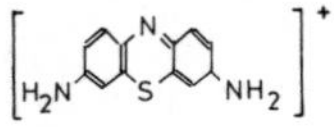

Fig. 7. Reaction schema for the photooxidation of Fe^{2+}-ions by thionine.

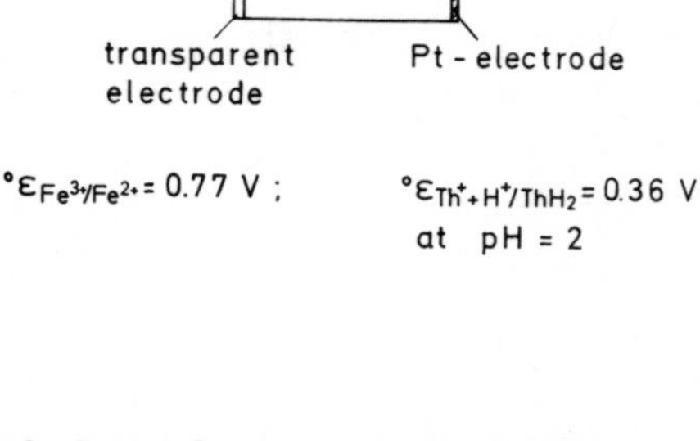

$$°\varepsilon_{Fe^{3+}/Fe^{2+}} = 0.77\ V\ ; \qquad °\varepsilon_{Th^+ + H^+/ThH_2} = 0.36\ V$$
at pH = 2

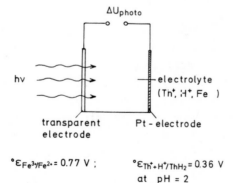

Fig. 8. Principle of photogalvanic cell with transparent counter electrode.

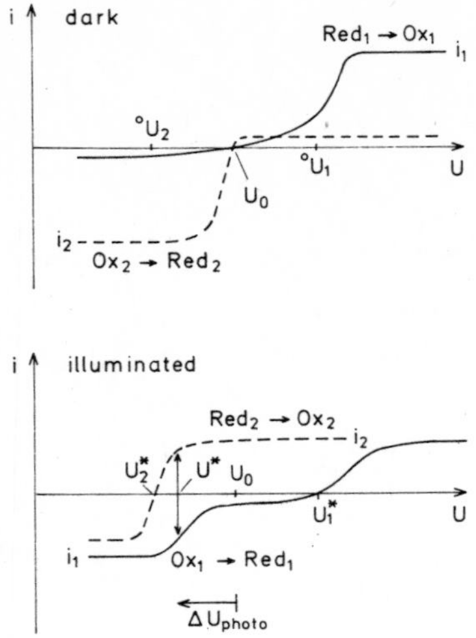

Fig. 9. Photogalvanic effect analyzed in terms of partial current voltage curves.

redox reactions will be shifted as a result of the new concentrations in this steady state, one in the cathodic and the other in the anodic direction. If the absolute increase in current for both processes is about the same as for the previous equilibrium potential, then the new steady state potential of the electrode will remain in the same range, and the photocurrents will be very small. Therefore, no photocurrents can be observed if both electrodes behave in the same way. The cell developed by Clark and Eckert [11] contains a transparent electrode composed of a doped SnO_2-glass. This electrode is selective for the thionine/leucothionine redox system and does not exchange electrons with the ferric-ferrous redox system to any appreciable extent. An electrode selective for the Fe^{2+}/Fe^{3+}-redox couple has not yet been found, and the obtainable photovoltage is therefore only about half of the theoretical limit. Since the reverse reaction in the solution cannot be prevented the quantum yield of such homogeneous photoreactions is low and the practical use for energy conversion seems to be rather limited.

Photoreactions of Excited Molecules at the Electrode-Electrolyte Interface

If molecules are adsorbed at an electrode surface, then the reaction partner of the excited molecule is normally the electrode itself. If the electrode has plenty of electron energy levels in the range of the energy levels of the excited molecule suitable for electron exchange, then electrons can easily be injected into the electrode or extracted from it. This is a normal situation for a metal electrode as is shown in Fig. 10. It is assumed here that the Fermi-level of the metal is located in the energy range between the two redox energy levels of the excited molecules.

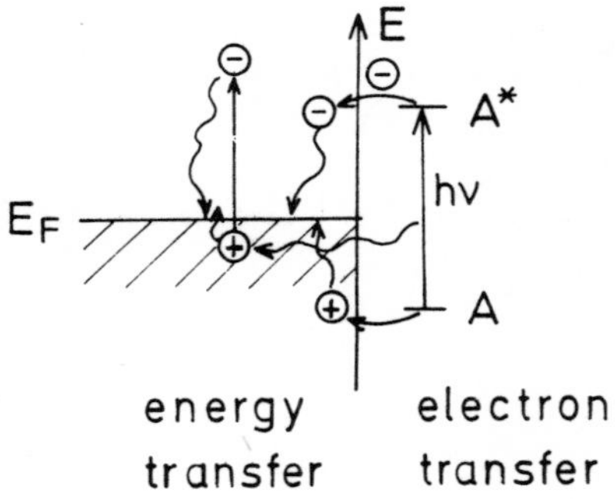

Fig. 10. Quenching mechanisms for excited molecule in contact with a metal by energy transfer and compensating electron transfer.

Since the molecule then reacts in both directions simultaneously no external current can be observed. It may happen that the process in one direction is much faster than in the other. But in this case the oxidized or reduced product will easily be restored to the original state by a reverse electron exchange with the metal. This is a well known mechanism of energy quenching by mutual exchange of electrons [13]. In parallel to this type of quenching, another very efficient energy dissipation mechanism involving energy transfer via dipole-dipole interaction (Förster mechanism) [14] will occur. Large effects can therefore only be expected if both quenching mechanisms can be excluded.

Just this situation is found at semiconductor electrodes. Energy transfer is not possible if the band gap of the semiconductor exceeds the excitation energy stored in the molecule. The reverse electron transfer is prevented if one of the redox levels of the excited molecule is located in an energy range within the band gap where no electronic states are available in the electrode. This is the reason why large photoeffects can be observed if excited molecules are adsorbed at semiconductors [3]. Fig. 11 gives a schematic picture of all possible correlations between the energy levels of an excited molecule and a solid electrode.

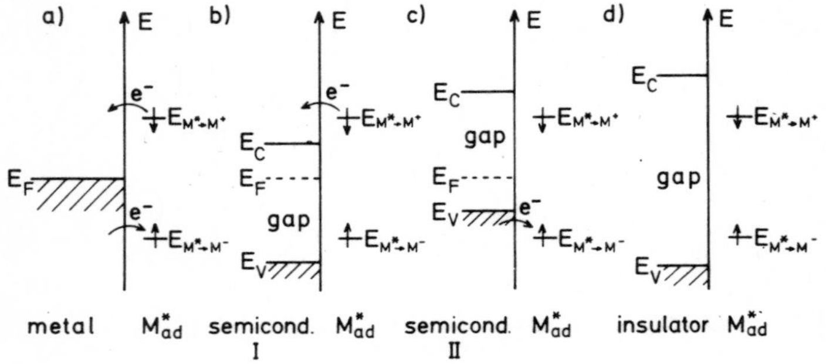

Fig. 11. Characteristic energy correlations between an excited molecule and the electronic energy states of various solids.

The three situations for semiconductors show that electron or hole injection is possible with the same excited molecule depending on whether the energy position of the conduction band or the valence band corresponds with one of the redox energy levels of the excited molecule. If the band gap is much wider than the excitation energy of the molecule, then no electron transfer can occur (right side of Fig. 11). If the injected electron or hole remains at the surface of the semiconductor, then it will, with a high probability, be recaptured by the mother-molecule. In studying such a sensitized charge injection it is necessary to apply a suitable voltage which removes the injected charge carriers from the surface. For electron injection one needs a positive, for hole injection a negative, voltage. A great number of studies have been done with zinc-oxide or other n-type materials and dyes which inject electrons into the conduction band [4,15-17]. The spectral dependence of the photocurrents at anodic bias reveals here very clearly the absorption spectrum of the excited dye molecules in the adsorbed state. The photocurrent spectra usually show a slight shift to longer wavelength's which is caused by the interaction with the substrate. Fig. 12 gives an example for the photocurrent spectrum of crystalviolet at zinc-oxide. The electrolyte in this experiment contained a redox system (quinone/hydroquinone) which acts as a regenerator for the oxidized dye molecules; otherwise the effect would decrease continuously with the extent of dye oxidation.

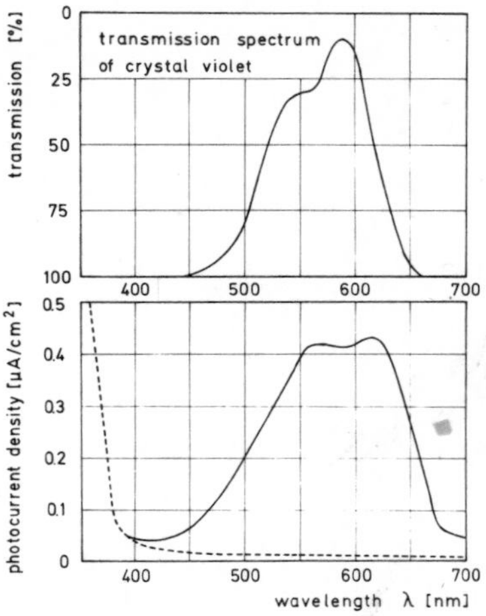

Fig. 12. Sensitized photocurrent spectrum due to crystalviolet adsorbed at a ZnO-electrode (electron injection). Upper part shows the absorption spectrum for comparison.

The same excited dye in contact with another semiconductor can inject holes and generate a cathodic photocurrent [4,18,19]. An example is shown in Fig. 13 where the photocurrent spectrum for crystal-violet adsorbed at a p-type gallium phosphide surface at cathodic bias is reproduced. Again the photocurrent spectrum corresponds to the absorption spectrum of the excited dye molecules.

If a redox system is present in solution which controls the electrode potential of the semiconductor and can act therefore like an external bias, photovoltages instead of photocurrents can be observed, which also reveal the absorption spectrum of the charge injecting dye. However, due to the non-linear correlation between photovoltages and the rate of charge injection, there is not a direct correspondence between the photovoltage and the absorption spectra. This is demonstrated in Fig. 14 where both the photocurrent spectrum and the photovoltage spectrum are shown.

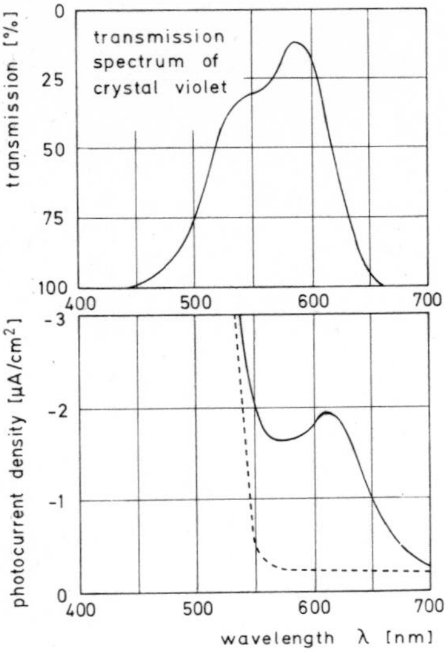

Fig. 13. Sensitized photocurrent spectrum due to crystalviolet adsorbed at GaP-electrode (hole injection) and absorption spectrum for comparison.

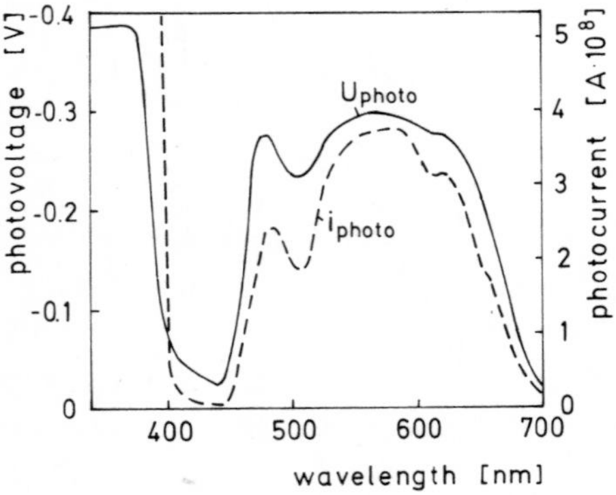

Fig. 14. Photovoltage and photocurrent in an electrochemical cell containing an illuminated ZnO-electrode in contact with adsorbed crystalviolet and a redox electrolyte with quinone/hydrochinone.

Only very small sensitized photovoltages or photocurrents have been found at silver halides where the electrochemical technique can be used in model studies of spectral sensitization in photography [20,21]. The efficiency is very low because the ionic conductivity of the silver halides severely limits the collection yield of injected electrons. An example for the photocurrent spectrum of a cyanine dye at silver bromide is shown in Fig. 15.

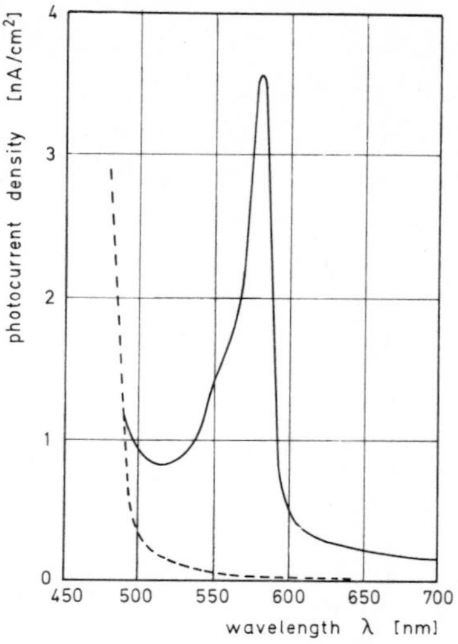

Fig. 15. Sensitized photocurrent at AgBr-membrane in contact with 1,1' diethyl-2,2' quinocyanin-chlorid as sensitizer.

Photoeffects Caused by the Excitation of the Electrode

We have seen in the introduction that the absorption of light in a solid generates excited electrons and excited vacant states, the so-called holes, which can cause photoelectrochemical reactions at the interface with an electrolyte (cf. Fig. 1). The magnitude of these effects depends on the lifetime of these carriers. We expect therefore large effects in semiconductors and eventually in insulators, and only small effects in metals. We shall discuss both types of electrodes because the effects in metals, despite their smallness, are of interest in principle. We shall, however begin with the discussion of semiconductor electrodes.

Photoeffects at Semiconductors

Apart from the long lifetime of the excited states in semiconductors, the electrostatic situation at the semiconductor-electrolyte interface can act in a very favourable way for reaching a high quantum yield. This is the case, if the semiconductor electrode is polarised to such a direction that an excess charge with a sign opposite to the majority carriers of the semiconductor is collected underneath the electrode surface. This gives a so-called depletion layer (depleted of mobile charge carriers) since the excess charge is constituted only of immobile ions [22]. The situation in the space-charge layer is usually represented in terms of the band scheme for electron energies which are dependent on the local electrostatic potential. This is shown in Fig. 16 for an n-type and a p-type semiconductor. The formation of a depletion layer at an n-type semiconductor needs a positive excess charge, at a p-type semiconductor a negative excess charge. The figure shows the energy of electrons versus the distance from the interface.

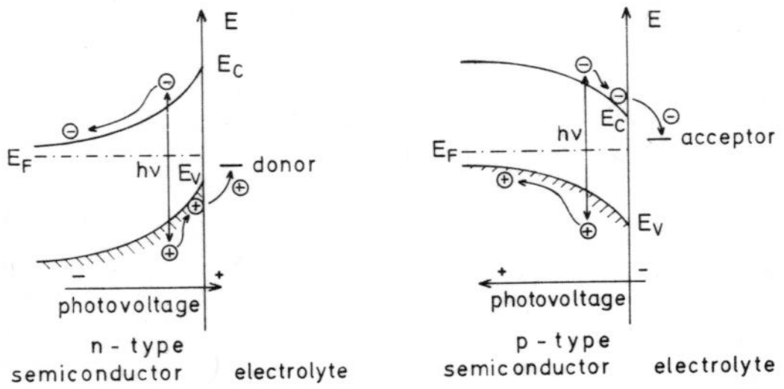

Fig. 16. Representation of a depletion layer at the semiconductor-electrolyte contact in terms of the electron energy band model (conduction band E_c and valence band E_v); origin of the photovoltage by charge separation.

Electrons in the conduction band have the tendency to move downwards in the energy scale, that means from the surface to the bulk in the n-type specimen, and the reverse in the p-type specimen. Holes in the valence band move upwards in this scheme, that is, to the surface in the n-type specimen and to the bulk in the p-type specimen.

This favourable situation for charge separation can be reached either by an external voltage, or by adding a suitable redox system to the electrolyte. An external voltage is necessary for studying photocurrents. A photovoltage is observed in an open cell if a depletion layer is formed in the dark, but its magnitude depends very much on the initial charge situation in the dark, since its maximum value under very high illumination intensity can only compensate the initial voltage drop in the space-charge layer. The origin of the photovoltage and its sign is also indicated in Fig. 16. The photocurrent spectra correspond to the absorption spectra of the semiconductor. They are, however, modified by the recombination efficiency which depends on the ratio between the penetration depth of the light and the extension of the space charge layer and on other factors. Some examples shall be given which demonstrate that these photoeffects can reach a quantum yield very close to one.

Fig. 17 shows the photovoltage and the photocurrent spectrum of an n-type cadmium sulfide electrode in a solution containing the redox system sulfide/polysulfide. The quantum yield for the photocurrent is of the order of 70%. The photoreaction at the electrode surface is the oxidation of sulfide ions by the incoming holes [23].

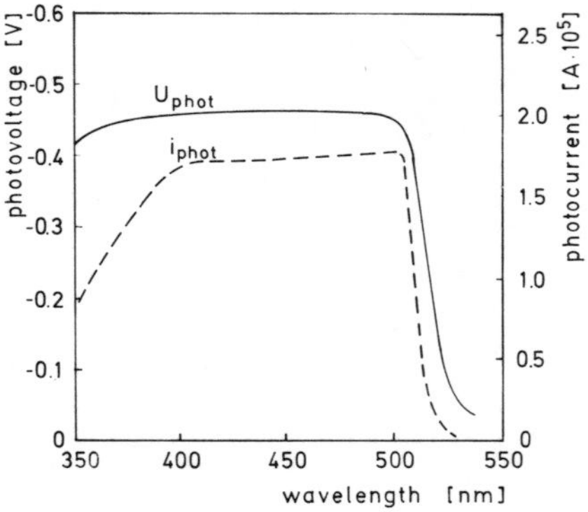

Fig. 17. Photovoltage and photocurrent spectra for n-type CdS electrode in Na_2S/Na_2S_x electrolyte.

Fig. 18 gives the equivalent photovoltage and photocurrent spectra of a p-type gallium phosphide electrode in the same redox system. This redox system implies here a negative excess charge on the gallium phosphide electrode causing a depletion layer of holes. The result is that the generated electrons move to the electrode surface and there reduce the polysulfide ions.

If no suitable redox couple is present in the electrolyte, the accumulation of holes or electrons in the surface of the semi-conductor usually leads to a decomposition of the semiconductor [24]. In this way zinc-oxide and cadmium-sulfide are photo-oxidized to form zinc ions and oxygen or cadmium ions and sulphur. p-type gallium phosphide is, however, not reduced because the hydrogen evolution prevents this reaction. Fig. 19 shows a

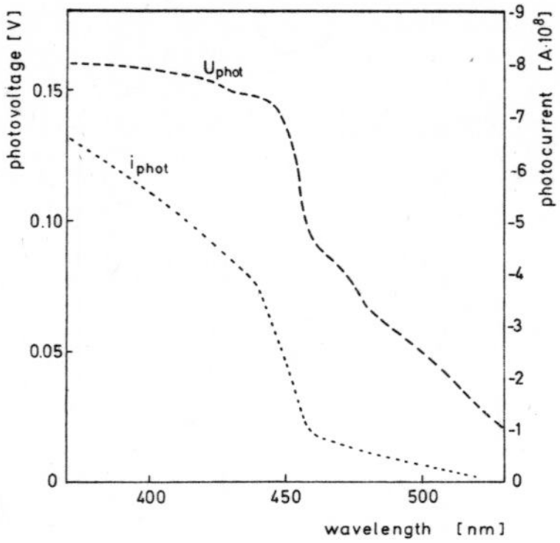

Fig. 18. Photovoltage and photocurrent spectra for p-type GaP-electrode in Na_2S/Na_2S_x electrolyte.

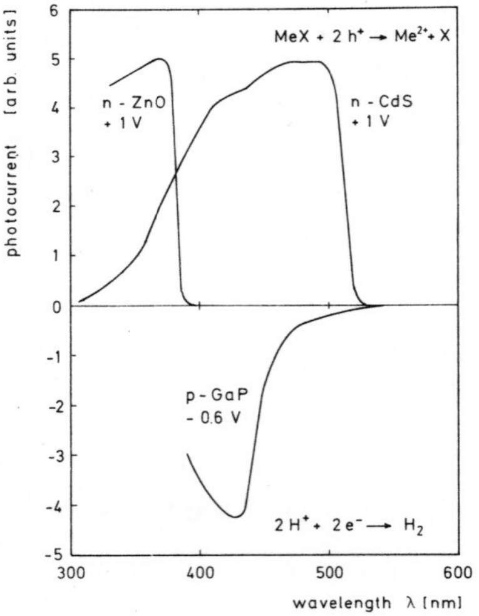

Fig. 19. Photocurrent spectra for n-type and p-type semiconductors in saturation region of the electrode potentials.

number of photocurrent spectra for various semiconductors which reveal the absorption spectra of the materials. The collection of reactions with electrons or holes in Fig. 20 give some examples of photodecomposition reactions caused by photoexcitation.

anodic oxidation of semiconductors

$$CdS + 2h^+ + aq \longrightarrow Cd^{2+}aq + S$$
$$ZnO + 2h^+ + aq \longrightarrow Zn^{2+}aq + \tfrac{1}{2}O_2$$
$$GaAs + 6h^+ + aq \longrightarrow Ga^{3+}aq + AsO_2^- + 4H^+ \cdot aq$$

cathodic reduction of semiconductors

$$CdS + 2e^- + aq \longrightarrow Cd + S^{2-}aq$$
$$ZnO + 2e^- + aq \longrightarrow Zn + 2OH^-$$
$$Cu_2O + 2e^- + aq \longrightarrow 2Cu + 2OH^-$$

Fig. 20. Table with some typical photodecomposition reactions of semiconductors in contact with electrolytes.

Photoeffects at Metal Electrodes

Photoelectron emission is a well known effect for illuminated metals in vacuum. If a metal is in contact with the liquid, the energy threshold for photoelectron emission can be either increased or decreased, depending on whether the interaction between the electrons and the other medium is attractive or repulsive. In aqueous solution we have an attractive interaction and the threshold for photoelectron emission is therefore decreased [25,26,27]. A comparison for the barrier between a metal and vacuum and a metal and an aqueous solution is shown in Fig. 21. This barrier represents the energy for free electrons in an aqueous solution where the motion of electrons is too fast to allow the orientation of water dipoles to follow. If an electron comes to rest and the water molecules have enough time to orient, then a solvated electron is formed [28,29] which has a lower energy, as is indicated in Fig. 21. This electron-dipole interaction plays no role in the process of electron emission; however, it stabilizes the emitted electrons in the solution.

Fig. 21 indicates that the energy position of the solvated electrons is still far above the Fermi-level of the electron-emitting metal. Since emitted electrons loose the excess kinetic energy very quickly in the sol-

ution and therefore reach only a very short distance from the electrode surface before they come to rest (the distance is in the order of 20-40 Å [25,
30]), the probability of being recaptured by the electrode is very high for
each emitted electron. It is therefore impossible to observe the photoelectron emission currents without preventing this recapturing process. The
presence of suitable electron scavengers can prevent this reverse electron
transfer reaction at the electrode [25,26,27]. Fig. 22 gives a scheme for

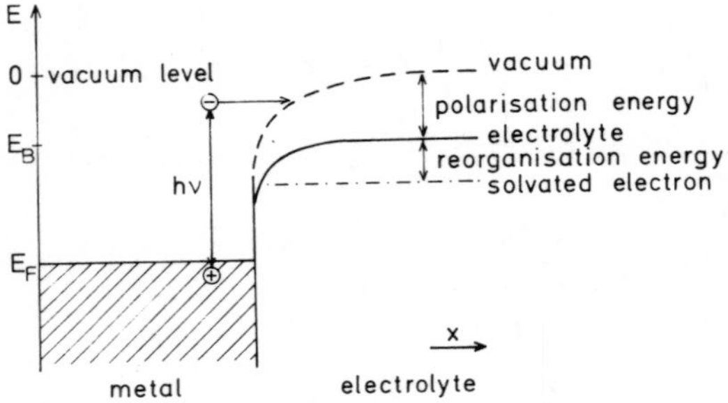

Fig. 21. Energy barrier for photoemission of electrons from metals into
electrolytes.

$$e^-_{excited} \longrightarrow e^-_{free} \longrightarrow e^-_{solvated}$$

$$e^-_{solvated} \xrightarrow{\text{diffusion}} e^-_{electrode}$$
$$\xrightarrow{\text{scavenger}} \text{products}$$

scavenger reactions :

$$N_2O + e^-_{solv.} \longrightarrow N_2 + O^-; \quad O^- + H_2O \longrightarrow OH^- + OH$$
$$OH + e^-_{electrode} \longrightarrow OH^-$$

$$H^+_{solv.} + e^-_{solv.} \longrightarrow H_{solv.}; \quad H_{solv.} + H_{ad} \longrightarrow H_2$$
$$H_{solv.} + H^+_{solv.} + e^-_{electr.} \longrightarrow H_2$$

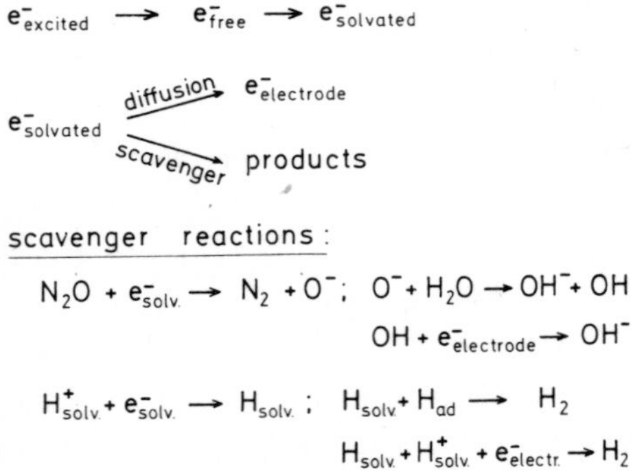

Fig. 22. Sequence of steps in photoelectron emission into electrolytes.

two useful scavenger reactions. N_2O and protons as scavengers have the
great advantage that their intermediate products can be more easily reduced
further rather than be reoxidized. This is indicated in the scheme of

Fig. 22. The advantage of such a consecutive electron transfer reaction between the electrode and the intermediates is that for each photoemitted electron another electron will be extracted, n this way doubling the observable current.

It has been found that the photoemission currents depend very much on the applied voltage. Fig. 23 gives such an example for two different electrode materials. The photocurrents have been measured here with the help of modulation techniques. The examples of Fig. 23 show that one observes not only a cathodic photocurrent, but also anodic photocurrents. These anodic photocurrents can be explained as reactions of excited holes in the metal [31].

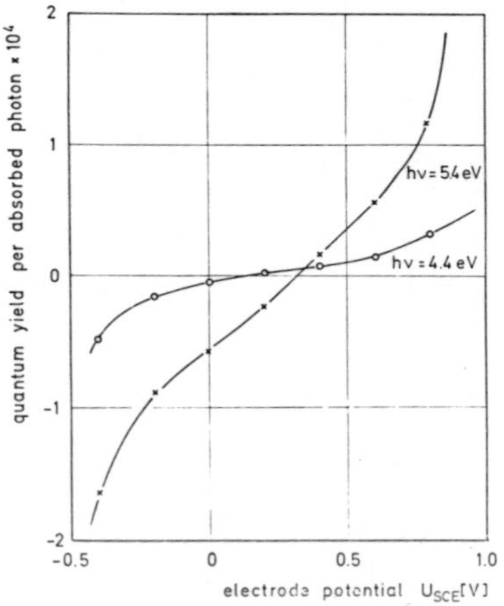

Fig. 23. Photocurrent-voltage curve for gold electrode in 1 n H_2SO_4, illuminated at two different wave lengths.

Light absorption in a metal generates not only excess electrons but also excited holes which have a high electron affinity like holes in semiconductors. The only difference is that the lifetime of these holes is much shorter in metals. However, if excited holes reach the interface between a metal and an electrolyte where electron donors are present, then these holes can oxidize the electron donors. In the case of a contact between a metal and an aqueous solution the water itself can act as an electron donor and is oxidized. Fig. 24 gives a representation in terms of electronic energy levels of such oxidation processes at illuminated metal surfaces in contact with an aqueous solution. It has been found that the photo-oxidatio in this case generates oxygen. A possible mechanism for this reaction is given in the scheme of Fig. 25 [31].

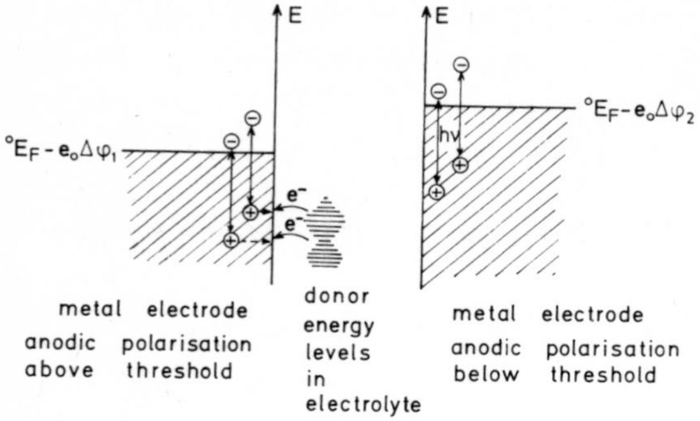

Fig. 24. Energy diagram to explain the reaction of excited holes in a metal with electron donors in the electrolyte and the influence of the electrode potential.

$$h^{+}_{excited} + H_2O \longrightarrow H^{+} + OH$$

$$OH + OH_{ad} \longrightarrow H_2O_2$$

$$H_2O_2 \longrightarrow O_2 + 2\,H^{+} + 2\,e^{-}$$

Fig. 25. Mechanism of reaction between excited holes and water.

We have seen in Fig. 23 that the cathodic as well as the anodic photocurrents depend on the applied voltage. The reason is that applying a voltage to a metal electrode immersed into a concentrated electrolyte solution varies primarily the potential drop in the Helmholtz double layer. This causes a variation of the potential barrier for photoelectron emission or photohole reactions in a very narrow range of the electrode-electrolyte interface.

As a consequence of varying the barrier height for photoelectron emission, the energy threshold can be changed continuously with the applied voltage. Increasing cathodic polarization reduces this barrier and increases therefore the photoemission yield, while polarization in the anodic direction decreases the photoemission current. For photohole reactions we have just the opposite influence of the electrode potential. The intersection point of the photocurrents with the abscissa in Fig. 23 means that at this potential anodic and cathodic photoeffects just compensate each other. Since both processes are independent reactions, it is understandable that this intersection point depends on the energy of the exciting photons. Fig. 26 demonstrates via an energy scheme the influence of a variation of the applied voltage on the excitation threshold for the two photoreactions.

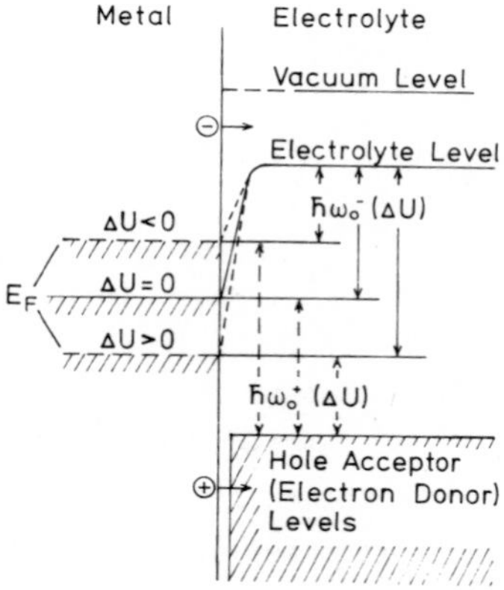

Fig. 26. Energy diagram to explain the potential dependence of cathodic and anodic photocurrents.

Finally, we shall give two examples for photocurrent spectra of the cathodic and the anodic photoeffects at metals. Fig. 27 shows how the quantum yield for photoemission currents at gold and copper electrodes depends on the energy of the exciting photon. The complicated energy dependence of the curves reveal the influence of the band structure on the emission process [32]. The "universal" law as postulated by some Russian authors [33,26] according to which the photocurrent should inc. ‿ase with the 5/2-power of the photon energy instead of the 2nd power as in vacuum photo-emission (Fowler's law [34]) seems only to be a special case for a

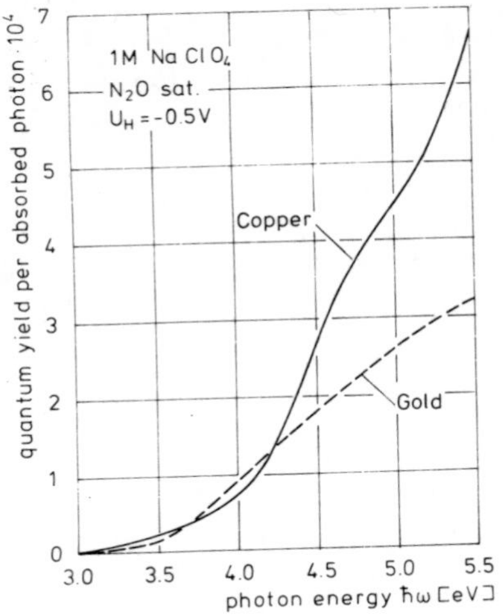

Fig. 27. Quantum yield of photoemission current of a gold and a copper electrode at the same electrode potential (equal threshold) in dependence on photon energy.

few metals. Fig. 28 shows the anodic photocurrents in dependence on the energy of the exciting photons for a gold electrode polarized to different voltages. The square root of the quantum yield is plotted because this should give a linear relationship between photocurrent yield and photon energy, if the Fowler relation for photoelectron emission can be applied. That this relation seems to be obeyed indicates that photohole reactions follow similar laws as photoelectron emission.

The study of photocurrents at metals can be used for various purposes One can get information about the excitation process in the solid, especially if one uses polarized light with varying angles of incidence as a tool for excitation [35]. On the other hand, photocurrents at metal electrodes can also give information on the structure of the electric double layer if more dilute electrolytes are used where teh barrier for photoemission is modified [36]. One can also get information on the position o the zero point of charge of a metal electrode [37], or of the energy level of free electrons in polar liquids [38].

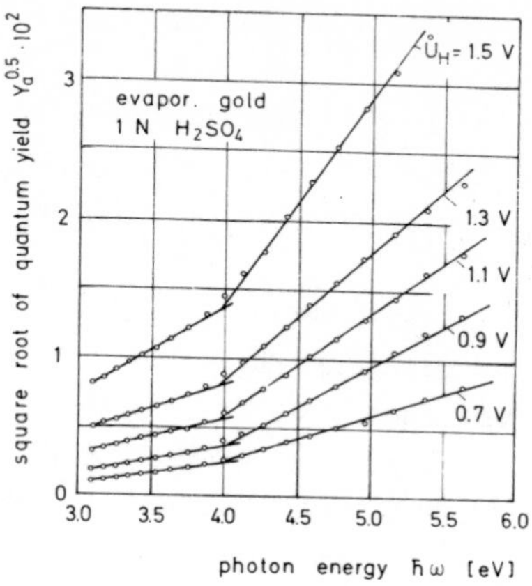

ig. 28. Quantum yield of anodic photocurrents at a gold electrode in de-
endence on photon energy at various electrode potentials.

In concluding it can be stated that the study of photovoltaic phenomena n electrochemical cells has opened a great number of new possibilities for he characterization of electronic processes and electronic energies in olecules, solids, and liquids.

EFERENCES

1. E. Becquerel, C. R. Acad. Sci. 9(1839)58, 145, 561.
2. cf. V. A. Myamlin, Yu. V. Pleskov, "Elctrochemistry of Semiconductors", Plenum Press, New York (1967).
3. cf. H. Gerischer, Ber. Bunsenges. Phys. Chem. 77(1973)771.
4. H. Gerischer, Faraday Discuss. 58(1974)219.
5. F. Lohmann, Z. Naturforsch. 22a(1967)813.
6. S. Trasatti, J. Electroanalyt. Chem. 52(1974)313.
7. cf. H. Gerischer and F. Willig, "Topics in Current Chemistry", 61(1976) 31.
8. H. Gerischer, Photochem. Photobiol. 16(1972)243.
9. E. Rabinowitch, J. Chem. Phys. 8(1940)551, 560.
. N. N. Lichtin: Internat. Conference on Photochemical Conversion and Storage of Solar Energy, London-Ontario 1976, J. J. Bolton (Editor), Wiley Publ. New York (1977), in press.

11. W. D. K. Clark and J. A. Eckert, Sol. Energy 17(1975)147.
12. M. D. Archer, J. Appl. Electrochem. 5(1975)17.
13. cf. M. Knibbe, D. Rehm and A. Weller, Ber. Bunsenges. Phys. Chem. 72 (1967)257.
14. Th. Förster, Ann. Physik 2(1948)55; Z. Elektrochem. 53(1949)93.
15. H. Gerischer and H. Tributsch, Ber. Bunsenges. Phys. Chem. 72(1968)42
16. K. Hauffe, V. Martinez, J. Range and R. Schmidt, Photograph. Korresp. 104(1968)427.
17. R. Memming, Faraday Discuss. 58(1974)262.
18. R. Memming and H. Tributsch, J. Phys. Chem. 75(1971)562.
19. R. Memming, Photochem. Photobiol. 16(1972)325.
20. H. Gerischer and H. Selzle, Electrochim. Acta 18(1973)799.
21. W. Jaenicke, Advances in Electrochem. and Electrochem. Engineering, H. Gerischer and Ch. W. Tobias (Editors), Vol. 10, J. Wiley Publ., New York (1977) p. 91.
22. cf. A. G. Milnes and D. L. Feucht, "Heterojunctions and Metal-Semiconductor Junctions", Academic Press, New York (1972).
23. A. B. Ellis, S. W. Kaiser and M. S. Wrighton, J. Am. Chem. Soc. 98 (1976)6855.
24. H. Gerischer and W. Mindt, Electrochim. Acta 13(1968)1509.
25. G. C. Barker, A. W. Gardner, and D. C. Sammon, J. Electrochem. Soc. 113(1966)1182.
26. A. M. Brodsky and Yu. V. Pleskov, "Progress in Surface Chemistry", S. G. Davison (Editor), Vol. 2, part 1, Pergamon Press, Oxford (1972)
27. Z. A. Rotenberg and Yu. Ya. Gurevich, J. Electroanal. Chem. 66(1975) 165.
28. cf. E. J. Hart and M. Anbar, "The Hydrated Electron", Wiley-Interscience, New York - London, (1970).
29. A. Henglein, Ber. Bunsenges. Phys. Chem. 78(1974)1078.
30. Z. A. Rotenberg, Elektrokhimiya 10(1974)1031.
31. H. Gerischer, E. Meyer and J. K. Sass, Ber. Bunsenges. Phys. Chem. 76 (1972)1191.
32. cf. "Photoemission from Surfaces", B. Fitton, B. Feuerbacher and R. F. Willis (Editors), Wiley-Interscience, New York (1977).
33. A. M. Brodsky, Yu. Ya. Gurevich and V. G. Levich, Phys. Stat. Sol. 4 (1970)139.
34. R. H. Fowler, Phys. Rev. 38(1931)45.
35. J. K. Sass, H. Laucht and K. L. Kliewer, Phys. Rev. Lett. 35(1975)1

36. A. M. Brodsky, Yu. Ya. Gurevich and S. V. Sherberstov, J. Electroanal. Chem. 32(1971)353.

37. Z. A. Rotenberg, Yu. A. Prishchepa and Yu. V. Pleskov, J. Electroanal. Chem. 56(1974)345.

38. R. R. Dogonadze, L. I. Krishtalik and Yu. V. Pleskov, Elektrokhimiya 10(1974)507.

ELECTROCHEMICAL SYNTHESIS

CHARLES K. MANN and MARGARET R. ASIRVATHAM
Department of Chemistry, Florida State University, Tallahassee, Florida 32306.

The field of electrosynthesis has been recognizable as an area of chemical specialization for more than a century. Its history has been marked by two major periods of activity, separated by a span of years. The first period developed from the fundamental work of Faraday and was spurred by the nineteenth century surge of activity in organic chemistry. The examination of electrochemical principles and reactions progressed steadily. We associate names such as Nernst, Kolbe, Haber and Tafel with these developments. It involved, for the most part, examinations of aqueous systems and reached a peak of activity in the period after World War I marked by the comprehensive works of Brockmann and Fichter.

At this point the rate of development diminished, probably owing to the restrictions imposed by the available equipment and solvent systems. With the developments in instrumentation that grew out of the technological advances for World War II and with the availability of relatively low cost nonaqueous solvents, conditions again became favorable for developments in this area, which continue to the present.

The modern period of activity has been marked first by a steady flow of new information to the original literature and then by publication of a series of reviews and monographs pertinent to the field. Those of general scope include the works of Swann [3], Allen [4], Mann and Barnes [5], Tomilov [6], Fry [7], Baizer [8], Weinberg [9] and Bard [10] which have appeared since 1956.

In addition to these which dealt primarily with electrochemical reactions, there have been a large number of publications in the field of electroanalytical chemistry that are significant in the context of this discussion because of contributions made by electroanalytical chemists to developments in instrumentation and methods of electrochemistry.

Some of the most prominent are the works of Kolthoff and Lingaine [11], Lingaine [12], Delahay [13], Adams [14] and Bard [15].

Electrochemical synthesis involves both inorganic and organic compounds. Some of the largest scale operations produce inorganics, e.g. aluminum and copper refining and the production of caustic and chlorine by electrolysis of brine. However these represent the products of a mature technology which has experienced only incremental development in recent years. The major thrust of recent published research has dealt with organic and organometallic electrochemistry and this discussion will emphasize these areas.

Useful chemical reactions may involve either breaking or forming chemical bonds; however reactions are more generally of synthetic interest when bonds are formed. The characteristic step of an electrochemical reaction is the transfer of an electron to or from a reactant by an electrode. More often than not, the electrode is effectively inert, so that the synthetic utility of electrochemical reactions mainly depends upon the nature of the chemical transformations which follow the initial electron transfer step. The actual electrode reaction therefore serves to produce reactive species which will ordinarily have limited lifetimes. Accordingly, the course of an electrochemical reaction is in a very major way dependent upon the environment that is provided the initial electrode product.

Most interest has centered upon reactions such as those involving ionic coupling, e.g. reductive coupling and polymerization of ions, on radical reactions such as the Kolbe synthesis and upon reactions involving nucleophilic attack upon electrochemically generated cation radicals. A study of electrosynthesis may therefore be undertaken by considering both the reactivities of molecules at electrodes and the strategies which have been developed for establishing environments which favor a subsequent useful transformation.

Introduction of an electron into a neutral, even-electron, molecule produces an anion radical. Depending upon its nature and that of the system, the species initially formed may or may not experience more than transient existence. The most commonly encountered cathodic reactions are illustrated in (EQ. 1-4).

$$RH + e^- \rightleftarrows RH^{\overline{\cdot}} \quad (\underline{I}) \tag{1}$$

$$\underline{I} + e^- \rightleftarrows RH^{-2} \rightarrow \text{Ionic Reactions} \tag{2}$$

$$\underline{I} + H^+ \rightarrow RH_2^{\cdot} \rightarrow \text{Radical Reactions} \tag{3}$$

$$RH_2^{\cdot} + e^- \rightarrow RH_2^- \rightarrow \text{Ionic Reactions} \tag{4}$$

The initial electron transfer product, I, may ordinarily be expected to add a proton to yield a neutral radical or to undergo decomposition so rapidly as to make its observation difficult or impossible. To achieve appreciable lifetimes for the initial products they generally must have structures which permit a high degree of stabilization by delocalizing the charge and spin density and they must be generated in systems from which acids and electrophilic reagents are excluded. Under these conditions, it may be possible to observe reversible oxidation and to recover the starting material, or occasionally, to form a dianion. Systems in which synthetically important reactions occur generally cause an immediate chemical reaction of the anion radical which may produce a species which also is reactive at the electrode. Cathodic reactions generally lead to acidic mixtures unless provision is made to control acidity.

In a similar manner, anodic oxidation of a neutral molecule produces a cation radical (EQ. 5) which may be expected to react as indicated in (EQ. 6-8). Cation radicals generally show more rapid reactions than do anion radicals and the immediate loss of a proton is the most probable second step.

If the reactions which have received attention from synthetic chemists are examined, many of them have the common attribute of requiring a small reaction potential, either nodic or cathodic, to initiate the initial electrode step. low reaction potential affords the possibility of effecting elective reaction by potential control and tends to favor the ormation of relatively less complex mixtures, compared to eactions which require higher potentials. Accordingly, in onsidering the scope and potential of electrochemical synhesis, it is appropriate to look at the ease of reactivity of arious kinds of compounds.

Variations in electrochemical reactivities, most ften expressed in terms of reaction potentials, tend to run arallel to variations in physico-chemical parameters such as cidities and frequencies of ultraviolet absorption maxima. hese analogies provide useful insight so long as the restriced scope of their applicability is borne in mind. Thus the rder of ease of reactivities of alkyl halides is I>Br>Cl>F, hich corresponds to the o ring of various chemical reactiviies. However, should alkyl amines be compared with benzene, heir relative frequencies of maximum ultraviolet absorption ould lead to the prediction that benzene should be more readly oxidized than amines, which is in fact not true. Thus orrelations of this type can be expected to hold only for omparison of quite similar compounds.

Saturated Compounds

As would be expected from other chemical and hysical properties, saturated molecules that do not contain toms with unshared pairs of electrons are very difficult to xidize or reduce. Reactions of these compounds appear to ave little synthetic utility. Actually, this is turned to dvantage, since it allows reactions to be run without causing eneral degradation of the molecule. Oxidation in aqueous ystems often causes attack at C-C and C-O bonds, converting he substrate to carbon dioxide and water. By contrast, eactions of saturated compounds that have atoms with unshared airs have been extensively investigated.

Alkyl Halides. The reactions of alkyl halides have eceived much attention since the pioneering investigations of on Stackelberg and Stracke [16]. After a period during which variety of mechanisms was advanced, there now seems to be general agreement that the mechanism initially presented by von Stackelberg (EQ. 9-11) applies to reduction of monohalides This shows the formation of hydrocarbon;

$$RX + e^- \rightarrow R\cdot + X^- \qquad (9)$$

$$R\cdot + e^- \rightarrow R^- \qquad (10)$$

$$R^- + HA \rightarrow RH + A^- \qquad (11)$$

however, in many cases the reaction does not go cleanly, but instead there are significant amounts of organometallics, olefins and products formed as a result of carbanion reactions Geminal polyhalides are more readily reduced than the corresponding monosubstituted compounds but the reaction proceeds via the same mechanism.

Vicinal dihalides also are reduced at less cathodic potentials than the analogous monohalide; however, the reaction is quite different. As shown in (EQ. 1-2), a two-electron process yields the olefin in a concerted reaction[17]

$$\begin{array}{c} X \\ | \\ >C-C< \\ \ \ \ | \\ \ \ \ X \end{array} + 2e^- \rightarrow \left[\begin{array}{c} X \\ \vdots \\ >C{\cdots}C< \\ \ \ \ \vdots \\ \ \ \ X \end{array} \right]^{-2} \rightarrow >C=C< + 2X^- \qquad (12)$$

Nonvicinal dihalides have been shown to undergo ring closure on reduction to 3- to 5-carbon rings (EQ. 13). The

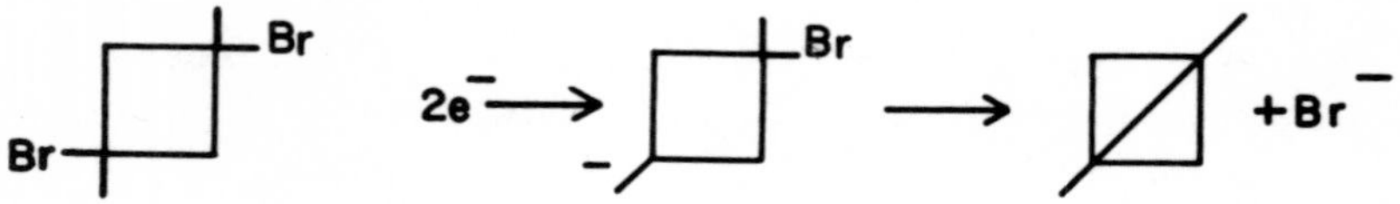

(13)

reaction, the steps of which are not concerted, appears to have synthetic utility, since quite good yields are possible [18].

Carbon-Nitrogen Compounds

Quaternary Amine Salts. Saturated quaternary amines
are not reduced at realizable potentials; instead their salts
are used as stable supporting electrolytes. However, if the
carbon-nitrogen bond is activated, e.g. by an alpha carbonyl
group, then the reduction takes place at modest potentials by
a reaction analogous to (EQ. 9-11) to give a hydrocarbon and
a tertiary amine [19].

This reaction has been made the basis for a synthesis of tertiary amines, using the sequence of steps illustrated in (EQ. 14-17) [20].

$$R_aN(CH_2C_6H_5)_2 + R_bBr \rightarrow R_aR_b\overset{+}{N}(CH_2C_6H_5)_2Br^- \qquad (14)$$
$$\underline{III}$$

$$\underline{III} + 2e^- + H^+ \rightarrow R_aR_bNCH_2C_6H_5 + CH_3Ph + Br^- \qquad (15)$$

$$R_aR_bNCH_2C_6H_5 + R_cBr \rightarrow R_aR_bR_c\overset{+}{N}CH_2C_6H_5Br^- \qquad (16)$$
$$\underline{IV}$$

$$\underline{IV} + 2e^- + H^+ \rightarrow R_aR_bR_cN + CH_3Ph + Br^- \qquad (17)$$

Saturated Amines. Aliphatic amines, not subject to
reduction, readily undergo oxidation. The reaction ordinarily
involves cleavage of a carbon-nitrogen bond to give the
less substituted amine, a reaction with no synthetic utility.
Substituted amines may show different reactions, however.
For example, substitution of a hydroxyl at the beta carbon
causes ring closure to form oxazolidines [21].

$$C_2H_5\overset{\overset{OH}{|}}{C}CH_2N(CH_3)_2 \longrightarrow \qquad (18)$$

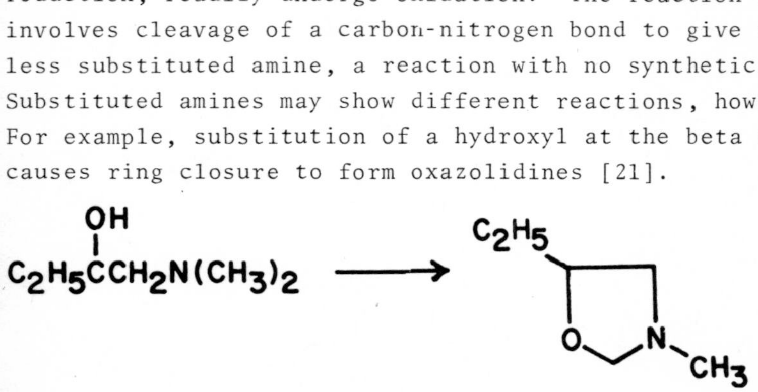

If the substituent positions of the carbon atoms alpha to the nitrogen are blocked, then the proton loss, analagous to (EQ. 7), cannot occur and the dealkylation process is inhibited. Thus, reaction of nortropane in benzonitrile produces the hydrazine VII [22].

$$OH^- \xrightarrow[-e^-]{CH_3CN} H_2O + \cdot CH_2CN \quad (19)$$

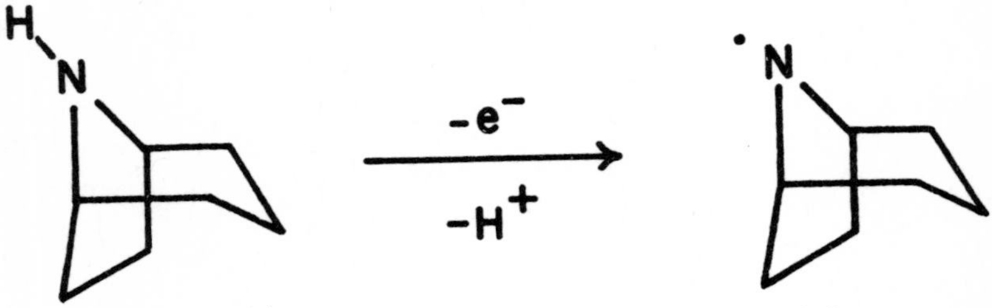

V

(20)

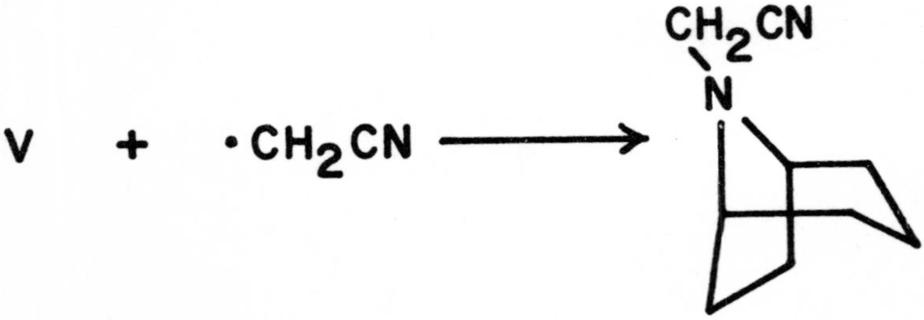

VI

(21)

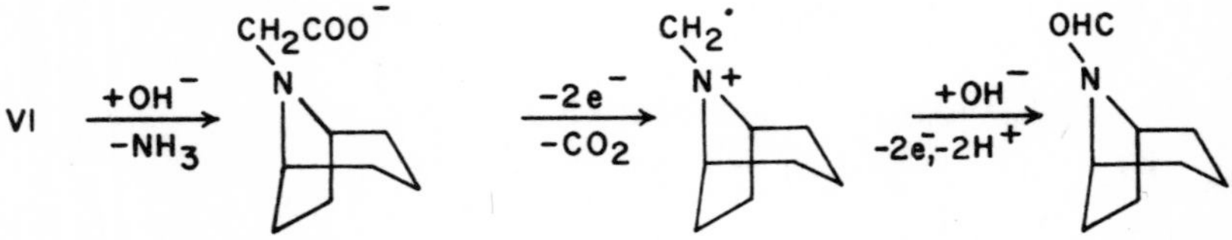

(22)

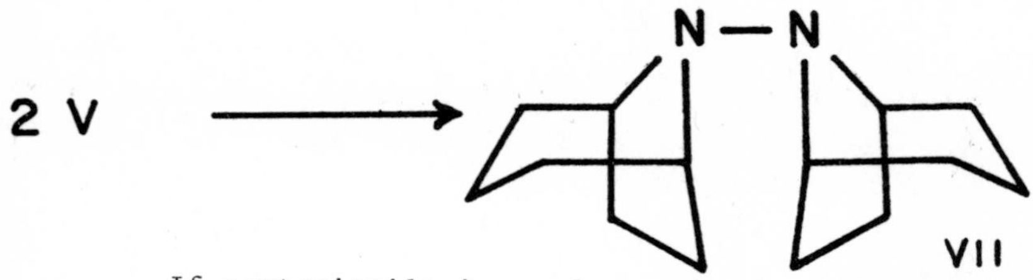

$$2 \, V \longrightarrow \quad \text{VII} \tag{23}$$

If acetonitrile is used as solvent with NaOH present, the major product is N-formylnortropane, which we believe to be caused by coupling of electrochemically generated cyanomethyl and amine radicals, followed by hydrolysis and additional oxidation.

Carbon-phosphorus Compounds. Unlike the analogous saturated ammonium salt, phosphonium salts, whether saturated or not, are subject to cleavage on reduction. In addition it occurs with retention of configuration. This has been made the basis for a synthesis of substituted phosphines, using a sequence of reaction analogous to (EQ. 14-17) [23, 24].

Unsaturated Compounds

Unconjugated unsaturated compounds, whether involving C=C, C=N, or C=O are difficult or impossible to oxidize or reduce and the reactions appear to have little synthetic utility. In conjugated systems, however, the situation is quite different and it is with them that the most important synthetic applications have been achieved.

Aromatic Nitrogen Compounds

Unlike aliphatic amines, compounds in which the nitrogen atom is conjugated with a pi system may undergo facile oxidation and reduction. In general, these reactions follow the schemes outlined in (EQ. 1-8), with the nature of the products determined by the chemical and electrochemical steps which follow the initial electron transfer. The oxidation of triphenylamine is shown in (EQ. 24) as an example [25].

$$2 \, Ph_3N \xrightarrow[-2H^+]{-2e^-} Ph_2N-\!\!\left\langle \bigcirc \right\rangle\!\!-\!\!\left\langle \bigcirc \right\rangle\!\!-NPh_2 \xrightarrow{-2e^-} Ph_2\overset{+}{N}-\!\!\left\langle \bigcirc \right\rangle\!\!-\!\!\left\langle \bigcirc \right\rangle\!\!-\overset{+}{N}Ph_2$$

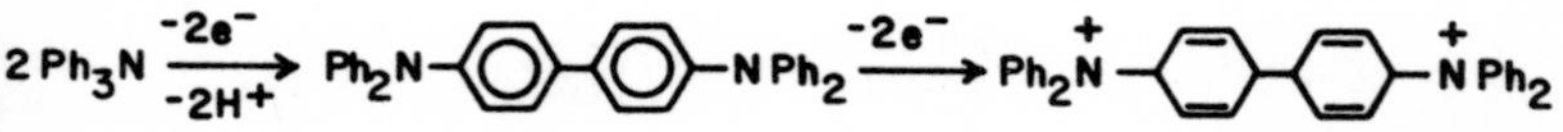

After the loss of an electron, the cation radical undergoes the loss of a proton and dimerizes to form the benzidine. This behavior is often observed with aromatic amines and phenols which do not have reactive sites blocked. In this case, the benzidine is oxidized at a less anodic potential than that required for triphenylamine, so the second electrode process occurs.

Reductive Polymerization. Cathodic reactions of olefins in which the double bond is conjugated with an activating group occur readily. For example, acrylonitrile is reduced in dilute solution to propionitrile (EQ. 25) through steps analogous to (EQ. 1,3,4). When the reaction is performed at high concentrations in systems which inhibit reaction by

$$CH_2{=}CH_2CN + 2e^- + 2H^+ \rightarrow CH_3CH_2CN \qquad (25)$$

water, adiponitrile is formed in good yields (EQ. 26). This is the basis for the successful process, developed by Manuel

$$2CH_2{=}CHCN + 2e^- + 2H^+ \rightarrow NC(CH_2)_4CN \qquad (26)$$

Baizer for the Monsanto Corp., which has been used for production of adiponitrile. It is also illustrative of a class of reductive coupling reactions, hydrodimerization, which is extremely versatile.

The reactions have been extensively investigated, especially by Baizer and coworkers, to evaluate their scope. In addition to activation by the cyano group, carboethoxy, diethylacetamido, pyridyl, diethoxyphosphonate, fluoroylidene and acetyl also allow formation of appreciable yields of hydrodimer. Mixed reductive coupling was shown to be feasible This type of reaction, involving two different activated olefins, can be expected to produce a mixture of three products if both compounds are reduced at the electrolysis potential. If there is at least 0.2 V difference in their otentials and if potential control is maintained to avoid the ore cathodic reaction, then the product should contain only he hydrodimer of the more easily reduced compound and the ixed product. The extent of cross coupling in this case will e affected by the relative tendencies of the two compounds o function as Michael acceptors. If compound A is more eadily reduced than B, and if B is present in excess, then the ross coupled product will be formed in good yields.

Various possible mechanisms have been suggested for eductive coupling. Of these, two are illustrated in (EQ. 27-1) in which X represents an activating group. In either equence, the initial step is addition of one electron to orm a radical anion IX (EQ. 27). One scheme involves eaction of IX with

$$CH_2{=}CHX + e^- \rightarrow \underset{(\underline{IX})}{[CH_2CHX]^{\bar{\cdot}}} \qquad (27)$$

$$\underline{IX} + CH_2{=}CHX \rightarrow \underset{(\underline{X})}{\left[\begin{array}{l} CH_2CHX \\ | \\ CH_2CHX \end{array}\right]^{\bar{\cdot}}} \qquad (28)$$

$$\underline{X} + 2H^+ + e^- \rightarrow \underset{(\underline{XI})}{X(CH_2)_4X} \qquad (29)$$

starting material, or with another acceptor for cross coupling, to produce a complex X which could be reduced to the inal product. Another possibility is the dimerization of IX o produce a dianion which could then give the product after rotonation (EQ. 30-31). Bard and coworkers have presented inetic evidence to show that this is the mechanism involved n the hydrodimerization of diethylfumarate [26].

$$2\ \underline{IX} \rightarrow \underset{(\underline{XII})}{X\bar{C}H(CH_2)_2\bar{C}HX} \qquad (30)$$

$$\underline{XII} = 2H^{+} \rightarrow \underline{XI} \qquad (31$$

The formation of a cyclic product from ethylcinnamate requires a slight modification of the mechanism above. Grypa and Maloy have suggested a rapidly established equilibrium between two radical anions and the dianion, which could undergo first order cyclization in the rate-determining step [27]. If the equilibrium favors the dianion, the reaction would be first order, otherwise it would be second order.

More recently, Nigretto and Bard have examined reactions at the second wave of diethylfumarate [28]. Their results show that the formation of dianion leads to dimers and trimers but not higher polymers. They suggested three possible reaction schemes which are outlined below. In these equations RH_2 indicates diethylfumarate.*

Proton transfer scheme

$$RH_2^{-2} + RH_2 \rightarrow RH_3^{-} + RH^{-}$$

$$RH^{-} + RH_2 \rightarrow R_2H_3^{-} \xrightarrow{+\ H^{+}} R_2H_4$$

$$R_2H_3^{-} + RH_2 \rightarrow R_3H_5^{-} \xrightarrow{+\ H^{+}} R_3H_6$$

Electron transfer scheme

$$RH_2^{-2} + RH_2 \rightarrow 2\ RH_2^{\bullet -}$$

$$2\ RH_2^{\bullet -} \rightarrow R_2H_4^{-2} \xrightarrow{+\ 2H^{+}} R_2H_6$$

---

*

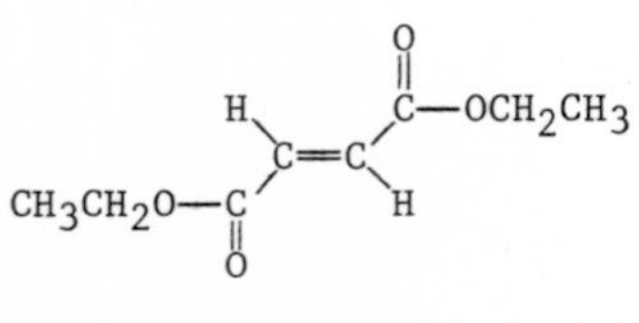

Direct addition scheme

$$RH_2^{-2} + RH_2 \rightarrow (RH_2)(RH_2)^{-2}$$

$$(RH_2)(RH_2)^{-2} + RH_2 \rightarrow (RH_2)_2RH_2^{-2} \xrightarrow{2\,H^+} R_3H_8$$

Oxidative Substitutions

Anodic substitution is another example of a general lass of reactions that has attracted interest because it ffers the potential for facile performance of an otherwise ifficult synthesis. This involves the anodic oxidation of a ixture of a substrate and reagent which can furnish the esired substituent. By this process it is possible to effect ntroduction to aromatic compounds of a variety of groups, uch as acetoxy, methoxy, acetamide, halide and cyano. ubstitution may be on the ring or on a side chain and by orrect adjustment of conditions it is often possible to nhance the yields of the desired product. This type of eaction has been studied by many investigators and detailed eviews are available [7-10].

The reaction generally involves oxidation of the ubstrate to a cation radical followed by proton loss, a econd electron transfer and by nucleophilic attack by the eagent. Acetoxylation of toluene is illustrated in (EQ. 32-3).

(32)

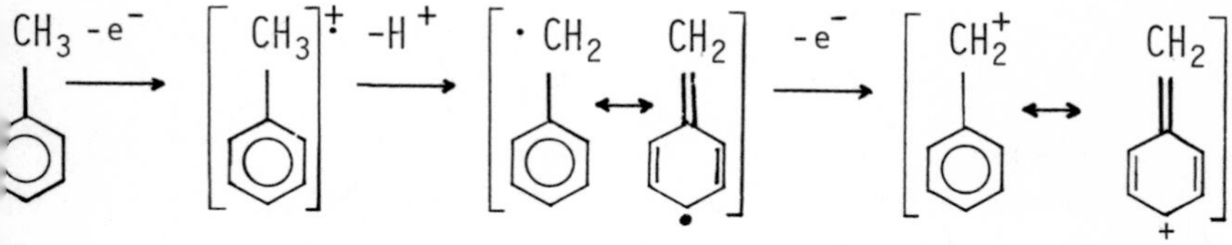

XIV

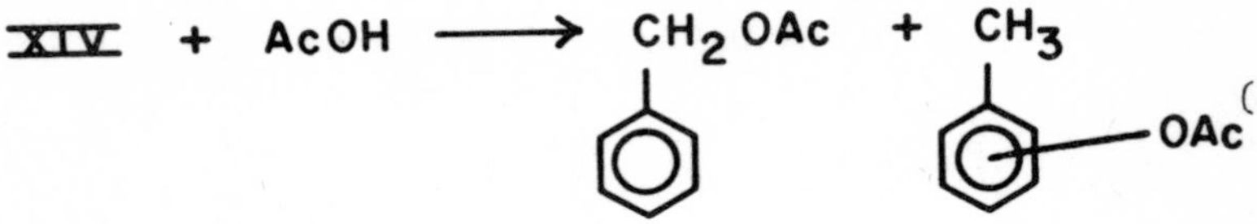

(33

In addition the product mixture may also contain dimers formed by reaction of XIII with the starting material. In some cases the nucleophilic attack results in addition, rather than substitution. An example, involving methoxylation, is shown in (EQ. 34) [29].

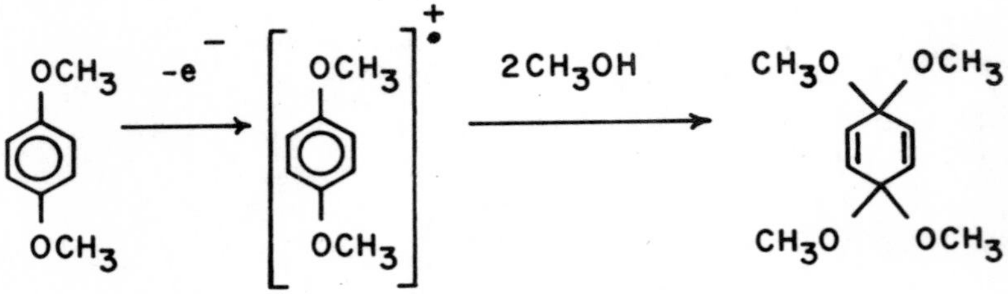

(34

Side chain substitution generally offers more synthetic promise because the reactions generally are cleaner and because it provides a procedure for introducing functiona groups into what might otherwise be an unreactive location. The side chain reaction will be enhanced by using a highly substituted substrate and by excluding nucleophilic agents from the reaction. If the substrate is itself a good nucleophile, dimerization will be encouraged.

Anodic Decarboxylation

Anodic decarboxylation has been the subject of extensive and continuing investigation for more than a centu The reaction of major interest involves oxidation of a carboxylate to produce a dimeric product (EQ. 35).

$$2\ RCOO^- - 2e^- \rightarrow RR + 2CO_2 \qquad (3$$

It is now generally agreed that the reaction involves discharge of the ionic starting material, followed by loss of carbon dioxide and the anodic oxidation of the resulting radical (EQ. 36). The reaction may be expected to produce a reasonable yield of dimer if the carboxylate is primary or if

$$RCOO^- - e^- \rightarrow RCOO\cdot \rightarrow CO_2 + R\cdot \rightarrow \tfrac{1}{2}RR \qquad (36)$$

it is substituted with electron-withdrawing groups. If the reactant is substituted with electron-donating groups, e.g. alkyl, alpha to the carboxyl, then the oxidation of the radical intermediate will be encouraged. The yield of dimer is reduced and products of ionic reactions, such as the alcohol in (EQ. 37) are formed.

$$R\cdot - e^- \rightarrow R^+ \xrightarrow{+OH^-} ROH \qquad (37)$$

The decarboxylation reaction can also be used to carry out electrochemically induced radical polymerizations as indicated in (EQ. 38-39).

$$^-O_2CRCO_2^- \xrightarrow[-CO_2]{-e^-} {}^-O_2CR\cdot \rightarrow {}^-O_2CRRCO_2^- \qquad (38)$$

(XV)

$$\underline{XV} \xrightarrow[-CO_2]{-e^-} {}^-O_2CRR\cdot \quad \text{etc.} \qquad (39)$$

It is interesting to note that when this reaction is run in aqueous or methanolic solutions, that the process of interest takes place at more anodic potentials than are required to discharge the solvent. The Kolbe reaction is then only possible because adsorption of the reactant on the electrode effectively inhibits solvent discharge, thus permitting the decarboxylation to occur. As is appropriate for a reaction that has received a great deal of attention over an extended period of time, excellent reviews are available[30-32].

Organometallic Compounds

The reactions and syntheses of organometallic compounds represent a striking exception to the general rule that, aside from metal refining and plating processes, the electrode serves as an electron source or sink and as a site and a catalyst for reactions of adsorbed intermediates, but is not a direct chemical participant. Many examples are known in which the electrode material reacts to form organometallics. These generally involve metals such as lead, mercury, tin and cadmium. They may involve either anodic or cathodic reactions. Many of these involve attack by an electrolytically generated intermediate on a metal cathode such as mercury and frequently constitute undesired side reactions. An example is shown in (EQ. 40-43) in which alkyl anions derived from ionization of a

$$RMgCl \rightleftarrows MgCl^{+} + R^{-} \tag{40}$$

$$4R^{-} + Pb - 4e^{-} \rightarrow R_4Pb \tag{41}$$

$$4MgCl^{+} + 4e^{-} \rightarrow 2Mg + 2MgCl_2 \tag{42}$$

$$Mg + RCl \rightarrow RMgCl \tag{43}$$

Grignard reagent are discharged at a lead anode. The cathode reaction continuously regenerates the anode reactant in this process which is used commercially for the production of tetra-alkyl leads [33]. An excellent detailed review of organometallic electrochemistry is given in Ch. 18 of reference 8.

Apparatus and Techniques

In addition to the apparatus and techniques which are generally used in synthetic chemistry, work in the area of synthetic electrochemistry requires the use of certain items which are peculiar to the field. Electrochemical techniques may be used either for the purpose of synthesis or to study reaction mechanisms. In either case, it is necessary to impose a sufficient potential across the interface between the working electrode and the solution to cause a reaction to occur. If the purpose of the reaction is investigative only, it may be possible to achieve satisfactory results by working on the scale of only a few tenths of an ampere or less. For large scale syntheses, the process will have to be scaled up. As is generally true of synthetic operations, this may involve specialized chemical engineering knowledge.

To carry out an experiment one must first decide upon the manner in which power is to be applied and upon the electrode system to be used. Reactions may be carried out either by controlling the reaction current or the potential. The technique of current control is simpler and is generally used when possible. It involves applying a sufficient voltage to a cell to cause the flow of a desired current, hence reaction at a defined rate. The utility of this approach is limited by the fact that it is not selective. Given a sufficient supply of potential reactants, the one most easily electrolyzed will react. However, if the rate at which the most reactive component is moved up to the electrode from the bulk of the solution is smaller than the reaction rate set by the imposed current, a mixed reaction will occur. Successful constant current operation requires either that the reactant of choice be maintained in adequate concentration, or that the competing reactions not cause difficulty in the subsequent workup. Economic operation obviously will only be possible if competing reactions are not large.

Operation at a controlled electrode potential makes possible a greater degree of selectivity, since a limiting reactant concentration at the working electrode can result simply in a limited current and not necessarily in a competing reaction. It will however, be necessary to arrange to control the potential at the electrode-solution interface by monitoring it with a reference electrode and adjusting the total cell voltage to maintain the desired electrode potential. Such an adjustment can be made manually to maintain control within rather broad limits of error. This is both imprecise and tedious and much better results can be obtained by using an automatic potentiostat.

In choosing electrodes, the aim is to select the most economical effective system. If the reaction to be run i a reduction, then a major factor is the hydrogen overvoltage a carbon and at most metals. Its discharge will greatly reduce the current efficiency of any competing reaction, accordingly, the available potential range for operation at high current efficiency will be restricted. The soft metals offer an extended working potential range because of their higher hydroge overvoltages. Mercury is especially useful because of the hig overvoltage of hydrogen and because cathodic reactions frequer tly proceed rapidly and cleanly at a mercury surface. If oxidations are to be run, then an anodically inert electrode is required. Those most frequently used are carbon, as graphite or one of the more compact forms, platinum and lead oxide

The counter electrode, e.g. the cathode in cells ir which the anodic reaction is of interest, must be one which does not produce an electrode product that complicates the workup procedure. This selection must be made in conjunction with that of the cell and supporting electrolyte. The cell design may provide for anode and cathode in the same or in separate compartments. Single compartment cells are advantageous in that electrical resistance can be minimized, thus reducing the power required for the process and simplifying tl cell design. They require very careful selection of the counter electrode since products formed at both electrodes will mix. On a laboratory scale, it may be feasible to use a relatively expensive electrode system in order to simplify the cell design. For example, a sacrificial silver anode may be used with a halide supporting electrolyte, assuring a solid electrode product in many solvents. For larger scale operations, more attention may be given to cell design in order to avoid the use of expensive components and reagents. Cells tha allow for flowing solutions are often used in large scale operation to minimize contact between anode and cathode products. A variety of membranes have been used to separate cel compartments. A detailed discussion of the design of cells a associated equipment has been presented by Goodrich and King Chapter 11 or Part I of reference 9.

Literature Cited

1. C.J. Brockmann, Electro-organic Chemistry, Wiley, New York 1926.

2. F. Fichter, Die Chemische Reaktion, Band VI, Organische Elektrochemie, Th. Steinkopff, ed. K.F. Bonhoeffer, Leipzig, 1942.

3. S. Swann, Technique of Organic Chemistry, Vol. II, 2nd Ed., A. Weissberger, ed., Interscience, New York, 1956, p.385.

4. M.J. Allen, Organic Electrode Processes, Reinhold Publishing Corp., New York, 1958.

5. C.K. Mann and K.K. Barnes, Electrochemical Reactions in Nonaqueous Systems, Dekker, New York, 1970.

6. A.P. Tomilov et. al., Electrochemistry of Organic Compounds, Israel Program for Scientific Translations, Ltd., Halstead Press, New York, 1972.

7. A.J. Fry, Synthetic Organic Electrochemistry, Harper and Row, New York, 1972.

8. Organic Electrochemistry, M.M. Baizer, ed., Dekker, New York, 1973.

9. Technique of Electroorganic Synthesis, N.L. Weinberg, ed., Wiley-Interscience, 1974.

10. Encyclopedia of Electrochemistry of the Elements, A.J.Bard, ed., a continuing series, Vol. 1-VI in 1976, Dekker, New York, 1976.

11. I.M. Kolthoff and J.J. Lingaine, Polarography, 2nd Ed., Interscience, New York, 1952.

12. P. Delahay, New Instrumental Methods in Electrochemistry Wiley, New York, 1954.

13. J.J. Lingaine, Electroanalytical Chemistry, 2nd Ed., Interscience, New York, 1958.

14. R.N. Adams, Electrochemistry at Solid Electrodes, Dekker 1969.

15. Electroanalytical Chemistry, A.J. Bard, ed., a continuing series, Vol. 1- , Dekker, New York.

16. M. von Stackelberg and W. Stracke, Z. Elektrochem., 53 (1949) 118.

17. J. Zavada, J. Krupicka and J. Sicher, Coll. Czech. Chem. Commun., 28 (1963) 1664.

18. M.R. Rifi, J. Amer. Chem. Soc., 89 (1967) 4442.

19. J.S. Mayell and A.J. Bard, J. Amer. Chem. Soc., 85(1963) 421.

20. L. Horner and A. Mintrup, Ann., 646 (1961) 49.

21. J. Klug and C.K. Mann, Unpublished data.

22. B.L. Laube, M. Asirvatham and C.K. Mann, J. Org. Chem., in press.

23. L. Horner and A. Mentrup, Ann., 646 (1961) 65

24. L. Horner et al. Tetrahedron Lett., (1961) 161.

25. D.W. Leedy and R.N. Adams, J. Amer. Chem. Soc. 92 (1970) 1646.

26. W.V. Childs, J.T. Maloy, C.P. Kesthelyi and A.J. Bard., J. Electrochem. Soc. 118, (1971) 874.

27. R.D. Grypa and J.T. Maloy, J. Electrochem. Soc., 122 (1975) 509.

28. J.M. Nigretto and A.J. Bard, J. Electrochem. Soc., 123 (1976) 1303.

29. B. Belleau and N.W. Weinberg, J. Amer. Chem. Soc., 85 (1963) 2525.

30. B.C.L. Weedon, Advan. Org. Chem., 1 (1960) 1.

31. A.K. Vijh and B.E. Conway, Chem. Rev., 67 (1967) 623.

32. L. Eberson in Chemistry of the Carboxyl Group, S. Patai ed., Wiley, New York, 1969, Ch. 6.

33. D.G. Braithwaite, J.S. D'Amico, P.L.Gross and W. Hanzel (to Nalco) U.S. Pat. 3287429 (1962); C.A.,66 (1967) 51716.

ECTROCHEMICAL CELLS WITHOUT LIQUID JUNCTION

TER A. ROCK

iversity of California, Davis, California

All operational electrochemical cells involve the physical separation reactive components in a manner that forces the reaction to proceed via e external circuit; an operational electrochemical cell must be free of ternal short circuits and spontaneous metathetical reactions. Some ex- ples of cells involving internal short circuits or spontaneous metathe- cal reactions are:

$(s)|PbSO_4(s)|CuSO_4(aq)|Cu(s)$ [1]

$(s)|K_4Fe(CN)_6(aq),K_3Fe(CN)_6(aq),KCl(aq)|AgCl(s)|Ag(s)$ [2]

$(s)|AgI(s)|HI(aq)|H_2(g)|Pt(s)$ [3]

$(s)|PbCrO_4(s)|K_2CrO_4(aq),KOH(aq)|HgO(s)|Hg(\ell)$ [4]

ll [1] is not operational because $CuSO_4(aq)$ is in direct contact with (s), and consequently the reaction

$$(s) + CuSO_4(aq) \rightarrow Cu(s) + PbSO_4(s)$$

oceeds spontaneously within the cell. Cell [2] is internally short cir- ited because both the oxidized and reduced forms of the redox couple $(CN)_6^{4-}/Fe(CN)_6^{3-}$ are in direct contact with the $Ag(s)|AgCl(s)$ electrode, d consequently, either the reaction

$$(CN)_6^{4-}(aq) + AgCl(s) \rightarrow Ag(s) + Cl^-(aq) + Fe(CN)_6^{3-}(aq)$$

the reverse reaction (depending on the concentrations of reactant and oduct solution species) is spontaneous within the cell. In fact it is possible to construct an operational cell involving a redox couple with luble oxidized and reduced forms that contains only a single electrolyte ase. Such cells are invariably short circuited in that they are equiva- nt to simply mixing the cell components together in a beaker. Cell [3] internally short circuited when the cell voltage is positive; the cell action is

$$(s) + HI(aq) = AgI(s) + \frac{1}{2}H_2(g)$$

cause HI(aq) is in direct contact with Ag(s) within the cell, the above reaction is spontaneous when $\varepsilon > 0$, that is, when the concentration of HI(aq) exceeds about 0.054 M (for P_{H_2} = 1 atm, 25°C). Because $H_2(g)$ is nc in direct contact with AgI(s), the reverse reaction is not spontaneous whe $\varepsilon > 0$. Cell [4] is not operational because the reaction

$$PbCrO_4(s) + 4OH^-(aq) \rightarrow CrO_4^{2-}(aq) + Pb(OH)_4^{2-}(aq)$$

occurs spontaneously within the cell. The $Pb(s)|PbCrO_4(s)|CrO_4^{2-}(aq)$ electrode is an example of an electrode of the second kind, $M(s)|MX(s)|X^{n-}$(sol where MX(s) is unstable in basic solution, and X^{n-} is unstable in acid solution ($Cr_2O_7^{2-}$ formation). Such electrodes, and redox couples of the type noted above, present special problems in the design of electrochemical cells without liquid junction.

Salt bridges often are employed in electrochemical cells to achieve a separation of reactive cell components. However, salt bridges are makeshift devices that give rise to liquid junction potentials of unknown magnitudes, and consequently salt bridges should be avoided in electrochemical thermodynamic investigations. Further, salt bridges obviously a undesirable in battery systems.

Although it is often assumed that cells without liquid junction cann be devised for certain types of electrodes, apparently it is always possible to devise an appropriate electrochemical cell without liquid junctio The primary limitation on the study of chemical reactions in electrochemic cells arises from electrokinetic factors, and not from spontaneous undesired reactions between cell components.

The variety of chemical reactions that can be investigated in electr chemical cells without liquid junction has been increased significantly b the realization that a cell with more than one electrolyte phase need not have a liquid junction, if a suitable ion-selective connector can be foun to separate physically the cell solutions [1,2,3]. The connector may be common central electrode (in which case the resulting cell is a double ce or an ion-selective membrane (either liquid or solid state).

A cell with liquid junction that is employed frequently in introducto discussions of electrochemical cells is the Daniell cell

$$Zn(s)|ZnSO_4(aq)||CuSO_4(aq)|Cu(s) \qquad [$$

where the double vertical bar denotes a liquid junction between the two cell electrolytes. The cell reaction of the Daniell cell is

$$Zn(s) + Cu^{2+}(aq) \rightleftharpoons Cu(s) + Zn^{2+}(aq) \pm \text{ion transfer} \qquad [$$

where "ion transfer" denotes the processes in which ions are transferred between the two cell electrolytes. It is these ion transfer processes that give rise to the liquid junction potential ε_J. Application of the Nernst equation to reaction [6] yields

$$\varepsilon = \varepsilon^\circ + \varepsilon_J - (RT/2F)\ell n(a_{Zn^{2+}}/a_{Cu^{2+}}) \simeq \varepsilon^\circ + \varepsilon_J - (RT/2F)\ell n[(Zn^{2+})/(Cu^{2+})] \quad [7]$$

where $\varepsilon_J = -\frac{RT}{2F}\int_L^R(t_{Zn^{2+}}d\ell na_{Zn^{2+}} + t_{Cu^{2+}}d\ell na_{Cu^{2+}} - t_{SO_4^{2-}}d\ell na_{SO_4^{2+}})$, and the t_i's are the transport numbers. The Daniell cell can be converted to a double cell without liquid junction by the use of common central mercury/mercurous sulfate electrode

$$Zn(s)|ZnSO_4(aq)|Hg_2SO_4(s)|Hg(\ell)|Hg_2SO_4(s)|CuSO_4(aq)|Cu(s) \quad [8]$$

The electrode reactions are

$$Zn(s) = Zn^{2+}(aq) + 2e^-$$

$$Hg_2SO_4(s) + 2e^- = 2Hg(\ell) + SO_4^{2-}(aq,L)$$

$$2Hg(\ell) + SO_4^{2-}(aq,R) = Hg_2SO_4(s) + 2e^-$$

$$Cu^{2+}(aq) + 2e^- = Cu(s)$$

and the net cell reaction is

$$Zn(s) + CuSO_4(aq) = Cu(s) + ZnSO_4(aq) \quad [9]$$

for which

$$\varepsilon = \varepsilon^\circ - (RT/2F)\ell n(a_{Zn^{2+}}a_{SO_4^{2-}}/a_{Cu^{2+}}a_{SO_4^{2-}}) \simeq \varepsilon^\circ - (RT/F)\ell n[(Zn^{2+})/(Cu^{2+})] \quad [10]$$

(Note the difference in the coefficient of the ℓn term in equations [7] and [10].) The operation of the common central electrode is thermodynamically equivalent to a semipermeable membrane that permits the transfer of SO_4^{2-} from one electrolyte solution to the other in a thermodynamically reversible manner. The net cell reaction [9] is much simpler than reaction [6], because the analysis of reaction [6] requires a detailed knowledge of ionic transport numbers and activities over a range of concentration for $ZnSO_4(aq) + CuSO_4(aq)$ mixtures.

Dilute-metal-amalgam connectors can be used in certain cases to join two electrodes of the second kind in a cell without liquid junction [4]. For example, the silver/silver ferrocyanide electrode can be investigated in the cell

$$Ag(s)|Ag_4Fe(CN)_6(s)|K_4Fe(CN)_6(aq)|\underset{\sim .01\%}{K(Hg)}|KCl(aq)|Hg_2Cl_2(s)|Hg(\ell) \qquad [11]$$

The electrode reactions of cell [11] are

$$4Ag(s) + Fe(CN)_6^{4-}(aq) = Ag_4Fe(CN)_6(s) + 4e^-$$

$$4K^+(aq,L) + 4e^- = 4K(Hg)$$

$$4K(Hg) = 4K^+(aq,R) + 4e^-$$

$$2Hg_2Cl_2(s) + 4e^- = 4Hg(\ell) + 4Cl^-(aq,R)$$

and the net cell reaction is

$$4Ag(s) + K_4Fe(CN)_6(aq) + 2Hg_2Cl_2(s) = Ag_4Fe(CN)_6(s) + 4Hg(\ell) + 4KCl(aq)$$

Note that K(Hg) does not appear in the net cell reaction and, consequently it is not necessary to know precisely the concentration of the potassium amalgam. The $Ag(s)|Ag_4Fe(CN)_6(s)|Fe(CN)_6^{4-}(aq)$ electrode is not operationa in appreciably acidic or basic solutions because of the instability of $Fe(CN)_6^{4-}(aq)$ and $Ag_4Fe(CN)_6(s)$, respectively. However, the use of a dilute-alkali-metal-amalgam connector in a double cell makes possible the study of this electrode in a cell without liquid junction.

Metal-amalgam connectors cannot be used for redox systems because the metal amalgam reacts with the oxidized form of the redox couple (internal short circuit). In such cases, the desired ion-selective characteristics of the central connector can be achieved by using ion-selective membranes as central connectors. For example, the anionic $Fe(CN)_6^{4-}(aq)/Fe(CN)_6^{3-}(aq)$ redox couple can be investigated [5] in the following cell without liquid junction

$$Au(s)|K_4Fe(CN)_6(m_2),K_3Fe(CN)_6(m_1)|K(memb)|KCl(m_3)|Hg_2Cl_2(s)|Hg(\ell) \qquad [12]$$

where K(memb) designates a potassium ion-selective membrane, for example, a K^+ ion-selective glass (27% Na_2O, 5% Al_2O_3, 68% SiO_2) or liquid (valinomycin in n-octanol) membrane. The net reaction of cell [12] is given by the sum of the following reactions

$$2Fe(CN)_6^{4-}(m_2) = 2Fe(CN)_6^{3-}(m_1) + 2e^-$$

$$6K^+(4m_2+3m_1) = 6K^+(4m_2+3m_1)$$

$$2K^+(4m_2+3m_1) = 2K^+(memb)$$

$$2K^+(memb) = 2K^+(m_3)$$

$$Hg_2Cl_2(s) + 2e^- = 2Hg(\ell) + 2Cl^-(m_3)$$

namely

$$2K_4Fe(CN)_6(m_2) + Hg_2Cl_2(s) = 2K_3Fe(CN)_6(m_1) + 2KCl(m_3) + 2Hg(\ell)$$

Data from cell [12] can thus be used to determine the value of $\varepsilon°$ for the ferro- ferricyanide couple.

Cationic redox couples, such as $Co(en)_3^{2+}(aq)/Co(en)_3^{3+}(aq)$ can be studied [6] in cells similar to cell [12], for example

$$Au(s)|Co(en)_3Cl_3(m_1),Co(en)_3Cl_2(m_2),NaCl(m_3),en(m)|Na(memb)|NaCl(m_4)|Hg_2Cl_2(s)|Hg(\ell) \quad [13]$$

where Na(memb) denotes a sodium ion selective glass (11% Na_2O, 18% Al_2O_3, 71% SiO_2) or liquid (e.g., a suitable crown ether in n-octanol) membrane. In this case NaCl must be added to the cell electrolyte containing the redox couple to insure the proper functioning of the Na(memb), as is shown in the following reaction sequence

$$Co(en)_3^{2+}(m_2) = Co(en)_3^{3+}(m_1) + e^-$$

$$3Cl^-(3m_1+2m_2+m_3) = 3Cl^-(3m_1+2m_2+m_3)$$

$$Na^+(m_3) = Na^+(memb)$$

$$Na^+(memb) = Na^+(m_4)$$

$$\tfrac{1}{2}Hg_2Cl_2(s) + e^- = Hg(\ell) + Cl^-(m_4)$$

The net cell reaction is

$$\{Co(en)_3^{2+}(m_2) + 2Cl^-(3m_1+2m_2+m_3)\} + \{Na^+(m_3) + Cl^-(3m_1+2m_2+m_3)\} + \tfrac{1}{2}Hg_2Cl_2(s)$$
$$= \{Co(en)_3^{3+}(m_1) + 3Cl^-(3m_1+2m_2+m_3)\} + Hg(\ell) + NaCl(m_4)$$

For both cells [12] and [13] the cation selective membrane connector does not appear in the net cell reaction, and therefore its properties, other than selectivity, need not be known in detail.

Solid-state ion-selective conductors are especially useful as connectors in high-temperature studies. For example, fluoride transfer reactions have been studied [7] in cells of the type

$$Fe(s)|FeF_2(s)|CaF_2(s)|NiF_2(s)|Ni(s) \quad [14]$$

Calcium fluoride is an ionic fluoride conductor at high temperatures (LaF_3 functions similarly at room temperature) and a piece of single CaF_2 crystal is used as the connector. The electrode reactions are

$$Fe(s) + 2F^-(CaF_2) = FeF_2(s) + 2e^-$$

$NiF_2(s) + 2e^- = Ni(s) + 2F^-(CaF_2)$

and the net cell reaction is

$Fe(s) + NiF_2(s) = FeF_2(s) + Ni(s)$

Yttruim-doped thoria has been used [8] as an oxide ion-selective conductor to study oxygen-transfer reactions. For example, the cell

$Pt(s)|CO(g),CO_2(g)|Y_2O_3(s),ThO_2(s)|Ga_2O_3(s)|Ga(\ell)$ [15]

has been used to investigate the high-temperature thermodynamic properties of $Ga_2O_3(s)$. The reaction of cell [15] is

$3CO(g) + Ga_2O_3(s) = 3CO_2(g) = 2Ga(\ell)$

Calcium stabilized zirconia can also be used in high-temperature cells to study oxygen-transfer reactions [9].

Sodium beta-alumina is used as a solid-state, sodium-ion selective separator in room temperature, high-energy-density battery prototypes [10]. For example

$Na(Hg)|Na_2O\cdot 11Al_2O_3(s)|NaOH(aq)|O_2(g)|Pt(s)$ [16]

for which the electrode reactions are

$2Na(Hg) = 2Na^+(Al_2O_3) + 2e^-$

$2Na^+(Al_2O_3) = 2Na^+(aq)$

$H_2O(\ell) + \frac{1}{2}O_2(g) + 2e^- = 2OH^-(aq)$

and the net cell reaction is

$2Na(Hg) + \frac{1}{2}O_2(g) + H_2O(\ell) = 2NaOH(aq)$

The need to employ cells without liquid junction is especially critical in electrochemical studies of isotope effects, because in such experiments the liquid junction potential may be of the same magnitude as the isotope effect itself. Electrochemical double cells without liquid junction have been employed successfully in the study of a variety of hydrogen and lithium isotope-exchange reactions [3]. For example, hydrogen-isotope-exchange reactions of the type

$D_2(g) + 2HX(soln) = H_2(g) + 2DX(soln)$ [17]

have been studied in cells of the type

$Pt(s)|D_2(g)|DX(soln)|MX(s)|M(Hg)|MX(s)|HX(soln)|H_2(g)|Pt(s)$ [18]

The net reaction for this cell is reaction [17]. Lithium-isotope-exchange eactions of the type

$$Li(s) + {}^{6}LiX(soln) = {}^{6}Li(s) + {}^{7}LiX(soln) \quad [19]$$

ave been studied in cells of the type

$$Li(s)|{}^{7}LiX(soln)|MX(s)|M(Hg)|MX(s)|{}^{6}LiX(soln)|{}^{6}Li(s) \quad [20]$$

ne net reaction for this cell is reaction [18]. Both cells [18] and [20] mploy the common central electrode connector of the second kind, as illusrated previously in more detail for cell [8].

The determination of Gibbs energies for solid-solid phase transitions nvolving metal salts generally requires very extensive thermal measurements n the separate phases. Direct study of solid-solid phase transition around oom temperature may require very high pressures (there is a unique pressure t each temperature at which the two phases are in equilibrium) and sophisicated experimental equipment. However, if both phases can be obtained at oom temperature and atmospheric pressure, then the Gibbs energy for the ransition can be determined in an electrochemical cell. For example, the ibbs energy of the aragonite-to-calcite phase transition has been deterined [11] in the cell

$$\underset{\text{-phase}}{\text{b(Hg)}}|PbCO_3(s),CaCO_3(\text{aragonite})|CaCl_2(aq,m_1)|Hg_2Cl_2(s)|Hg(\ell)\ -$$

$$Hg(\ell)|Hg_2Cl_2(s)|CaCl_2(aq,m_1)|CaCO_3(\text{calcite}),PbCO_3(s)|\underset{\text{2-phase}}{Pb(Hg)} \quad [21]$$

ne electrode reactions are

$$b(Hg) + CaCO_3(\text{aragonite}) = PbCO_3(s) + Ca^{2+}(aq) + 2e^-$$

$$g_2Cl_2(s) + 2e^- = 2Hg(\ell) + 2Cl^-(aq)$$

$$Hg(\ell) + 2Cl^-(aq) = Hg_2Cl_2(s) + 2e^-$$

$$bCO_3(s) + Ca^{2+}(aq) + 2e^- = CaCO_3(\text{calcite}) + Pb(Hg)$$

f the $CaCl_2(aq)$ solution is the same in both halves of the cell, then the um of the above electrode reactions yields as the net cell reaction

$$aCO_3(\text{aragonite}) = CaCO_3(\text{calcite})$$

ne cell voltage is independent of the $CaCl_2(aq)$ concentration, $\varepsilon_{cell} = \varepsilon^\circ_{tr}$, nd $\Delta G^\circ_{tr} = -2F\varepsilon^\circ_{tr}$.

Equilibrium constants for complexation reactions also can be determined eadily by the double-cell method. For example, the reaction

$${}^{2+}(aq) + X(s) \rightleftharpoons BaX^{2+}(aq)$$

Where X is the crown ether dicyclohexyl-18-crown-6

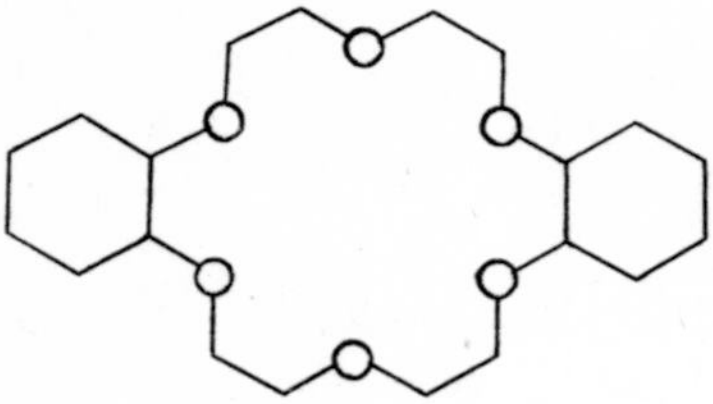

has been studied [12] in the cell

$$\underset{\text{2-phase}}{Pb(Hg)}|PbCO_3(s),BaCO_3(s)|BaXCl_2(aq,m_1),X(s)|Hg_2Cl_2(s)|Hg(\ell) -$$

$$- Hg(\ell)|Hg_2Cl_2(s)|BaCl_2(aq,m_1)|BaCO_3(s),PbCO_3(s)|\underset{\text{2-phase}}{Pb(Hg)} \qquad [22]$$

The electrode reactions are

$$Pb(Hg) + BaCO_3(s) + X(s) = BaX^{2+}(aq) + PbCO_3(s) + 2e^-$$

$$Hg_2Cl_2(s) + 2e^- = 2Hg(\ell) + 2Cl^-(aq)$$

$$2Hg(\ell) + 2Cl^-(aq) = Hg_2Cl_2(s) + 2e^-$$

$$Ba^{2+}(aq) + PbCO_3(s) + 2e^- = BaCO_3(s) + Pb(Hg)$$

The left cell electrolyte is from the same $BaCl_2$(aq) stock solution as the right electrolyte, except that it is saturated with the complexing agent X. Under these conditions, the net cell reaction, obtained by summing the above electrode reactions is

$$BaCl_2(aq) + X(s) = BaXCl_2(aq) \qquad [23]$$

If the activity coefficients of $BaCl_2$ and $BaXCl_2$ are approximately equal at equal concentrations, then $\varepsilon_{cell} = \varepsilon^\circ_{23}$.

The use of central connectors makes possible the study of a wide variety of interesting chemical reactions in cells without liquid junction, and greatly increases the scope of electrochemical cell design.

REFERENCES

1. P.A. Rock, J. Chem. Educ., 47(1970)683.
2. P.A. Rock, J. Chem. Educ., 52(1975)787.
3. P.A. Rock, ACS Symp. Ser. V. 11, Amer. Chem. Soc., Washington, D.C., 1975, p. 131.
4. P.A. Rock and R.E. Powell, Inorg. Chem., 3(1964)1593.
5. R.C. Murray, Jr., and P.A. Rock, Electrochim. Acta, 13(1968)969.

6. J.J. Kim and P.A. Rock, Inorg. Chem., 8(1969)563.
7. G. Chattopadhyay, M.D. Karkhanavala and M.S. Chandrasekharaiah, J. Electrochem. Soc., 122(1975)325.
8. K.A. Klinedinst and D.A. Stevenson, J. Chem. Thermo., 4(1972)565 and 5(1973)21.
9. E.S. Ramakrishnan, O.M. Sreedharan and M.S. Chandrasekhariah, J. Electrochem. Soc., 122(1975)328.
0. F.G. Will and S.P. Mitoff, J. Electrochem. Soc., 122(1975)457.
1. P.A. Rock and A.Z. Gordon, J. Amer. Chem. Soc., 98(1976)2364.
2. A.Z. Gordon and P.A. Rock, J. Electrochem. Soc. 124(534)1977.

SPECIES-SELECTIVE ELECTROCHEMICAL SENSORS

PETER A. ROCK
Department of Chemistry, University of California, Davis, California.

INTRODUCTION

The scope and utility of electrochemical measurements has been expanded enormously during the last decade by the development of new types of species-selective electrochemical sensors [1,2,3,4]. Electrochemical sensors are available for the measurement, even in complex mixtures, of the activity of a wide variety of chemical species, including inorganic and organic ions and neutral molecules.

The basic problem involved in the development of a species-selective electrode is to find a semi-permeable membrane (liquid, solid, or glass) that (ideally) is permeable (a) only to the species of interest, or (b) that is permeable only to a product of a reaction of the species of interest that occurs adjacent to the external surface of the semi-permeable membrane. The semi-permeable membrane is used to separate the metal-electrolyte interface, where the electron transfer processes occur, from the solution containing the species of interest; the membrane prevents other species in the solution from interfering with ("poisoning") the electron transfer processes.

The principal types of semi-permeable membranes that are used in species-selective electrodes are:

1. Liquid-ion-exchange membranes. This type of membrane involves a mobile ion-carrier (ion-complexing agent) dissolved in a solvent that is immiscible with the solution of interest. The complexing agent forms a complex with the ion of interest and thereby provides a mechanism for tha ion to get through the membrane. Ions that do not form a stable complex with the carrier cannot pass through the membrane.

2. Solid-state membranes. This type of membrane is based on the fa that certain crystals and glasses exhibit ionic conductivity, that is, th ions in certain solids and glasses can move through the lattice network b utilizing vacancy or interstitial sites.

3. Heterogeneous (or Pungor-type) membranes. In this type of membrane, insoluble salts, usually involving the ion of interest, are imbedd

in a rubber (usually silicone) matrix. The membrane composition is typically 50-60% (by wt.) salt. Transport mechanisms vary but in any case presumably involve transfer of ions between crystals that are in physical contact within the matrix.

4. Polymer membranes. This type of membrane is employed in several varieties of gas-sensing electrodes. Transport involves the passage of small neutral molecules through openings in the polymer network.

5. Multilayered membranes. This type of membrane consists of one of the above types of membrane that is covered on the side contacting the solution of interest with a layer containing, for example, an enzyme that is specific for the species of interest. The underlying semipermeable membrane is selective for a specific product of the enzyme reaction.

General Considerations

Before proceeding to a consideration of specific examples of ion-selective electrodes, it is desirable to outline several general aspects of ion-selective electrode behavior.

1. Electrodes respond to activity and not to concentration. ($a_i = \gamma_i c_i$, where a_i is the activity of species i, c_i is the concentration, and γ_i is the activity coefficient). In addition, electrodes respond to free (i.e., uncomplexed) ions; the electrode does not measure that portion of the total amount of the ion that is in the complexed form. For example, in a solution containing $Ca^{2+}(aq)$ and $CaX^{2+}(aq)$, where X is a complexing agent, a calcium-ion-selective electrode measures only the activity of $Ca^{2+}(aq)$; the concentration of free Ca^{2+} may be only a very small fraction of the total calcium.

2. No ion-selective electrode is perfectly selective. The imperfect selectivity of ion-selective electrodes arises because more than one species may be able to utilize the same transport mechanism. Ions with the same charge as the ion of interest that also have a similar size are strong candidates as possible interferants. In any case, possible or known interferants in the solution must be carefully considered and appropriate selectivity tests should be carried out.

The modified Nernst Equation for an ion-selective electrode that responds to two or more ions is [6]

$$E = E° + E_J - \frac{RT}{z_i F}\ln(a_i + \sum_j k_{ij} a_j^{z_i/z_j}) \qquad [1]$$

where a_i is the activity of the ion i, which is the ion of interest, and the a_j values are the activities of the various interferant ions to which the electrode also responds. The k_{ij} values are the experimentally determined [6] selectivity coefficients; z_i and z_j are the ion charges; and, E_j is the liquid junction potential, which arises if the measurement cell involves a liquid junction. The other symbols have their usual meanings. I order for an ion-selective electrode to be useful in monitoring the activity of ion i, it is necessary that $k_{ij}a_j^{z_i/z_j} \ll a_i$ for all j ions in th solution, or that k_{ij} be known for any j ion for which the above inequalit does not hold. An effective ion-selective electrode is one for which $k_{ij} \simeq 0$ for all j ions that commonly occur in solutions where the i ion is to be monitored (e.g. K^+ with Na^+, or Mg^{2+} with Ca^{2+} in biological fluids)

In addition to interferences of the type considered above, it is also neces sary to consider possible interferences arising from reactions between the components of the solution of interest and (a) the components of the semi-permeable membrane of the ion-selective electrode, or (b) the components of the reference electrode. In measurements on systems containing proteins or colloids, consideration must be given to the possible adsorption of proteins or colloids on the semi-permeable membrane, or on the frit junctions of the reference electrode (Pallman Effect). The Pallman effect can give rise to very large errors in a_i.

3. Measurement errors. Measurements of moderate-to-high precision require a high-quality electrometer or digital voltmeter with a very high input resistance. A 1 mV error in the measured voltage corresponds to an 8% error in (M^{2+}), a 4% error in (M^+), and a 2% error in the pH:

Precision in E(mV)	% error in (M^+)*	% error in (M^{2+})*
0.1	0.4	0.8
1.0	3.8	7.5
2.0	7.5	15.0

* $\Delta E \simeq \frac{60}{z} \log \left[(M^{2+})'/(M^{2+})\right]$

Measurements involving one ion-selective electrode versus another require the use of an integrated circuit differential amplifier [7].

In work aspiring to high accuracy, especially in measurements designe to yield thermodynamic data, the measurement system should be free of liquid junctions [8]. The use of cells without liquid junction avoids the problem of single ion activities [9].

Ion-specific electrodes should be calibrated frequently (and preferably just before use), because $E°$ is a property of the particular electrode and in some cases $E°$ changes with time. In cells involving liquid junctions, the calibration solutions and the solutions to be analyzed should have similar ionic strengths and compositions in order to minimize variations in E_J.

Advantages of Ion-Specific Electrodes. The principal advantages of ion-specific electrodes are: accuracy; reliability; convenience (easy to use and adaptable, e.g., they can be set up for use in controlled atmospheres); continuous and unattended monitoring; selectivity (can monitor a specific species in a complex mixture for which the detailed composition need not be known); miniaturization (for biological studies); and, non-destructive sample analysis. Multi-species probes can be constructed and placed in relatively inaccessible locations (e.g., body organs, flue stacks and exhaust pipes, sewer systems, etc.).

Representative Systems

I. Liquid Ion-Exchange Membrane Systems

A cross-section of a calcium ion-selective liquid-ion-exchange membrane electrode is shown below:

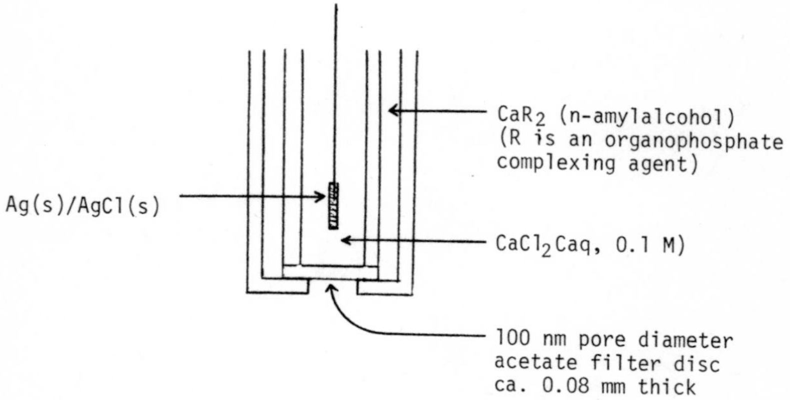

ig. 1. Cross-sectional view of a calcium liquid ion-exchange electrode.

The electrode processes are

$$2Ag(s) + 2Cl^-(0.2\ M, aq) \longrightarrow 2AgCl(s) + 2e^-$$

$$Ca^{2+}(0.1\ M, aq) + 2R^-(alc) \longrightarrow CaR_2(alc)$$

$$CaR_2(alc) \longrightarrow Ca^{2+}(aq, ext) + 2R^-(alc).$$

The sum of the above electrode processes yields

$$2Ag(s) + CaCl_2(aq, 0.1\ M) \longrightarrow 2AgCl(s) + Ca^{2+}(aq, ext) + 2e^-. \qquad [2]$$

Note that the properties of the liquid ion-exchanges do not appear in the net cell reaction, and therefore these properties need not be known in detail for thermodynamic purposes. Note also that the activities of all the species in reaction [2] are constant for a given electrode at a particular temperature. Consequently, we can write

$$E = E° - \frac{RT}{2F}\ell n a_{Ca^{2+}}.$$

Reaction [2] is formally equivalent ot the reaction

$$Ca(s) \longrightarrow Ca^{2+}(aq, ext) + 2e^- \qquad [3]$$

for which $E = E° - \frac{RT}{2F}\ell n a_{Ca^{2+}}$. The $Ca(s)/Ca^{2+}(aq)$ electrode is not operational because of the spontaneous reaction of Ca(s) with water.

The criteria for the solvent of the liquid ion-exchanger are: low-vapor-pressure, low-dielectric-constant, high-molecular-weight organic liquid that is immiscible with water (e.g., n-amylalcohol or n-octanol). The millipore filter disc is treated to make it hydrophobic (silicone fluid), and the organic liquid wicks into the pores of the filter from the reservoir. The R^- group in the ion-exchanger is an organic phosphate diester, for example

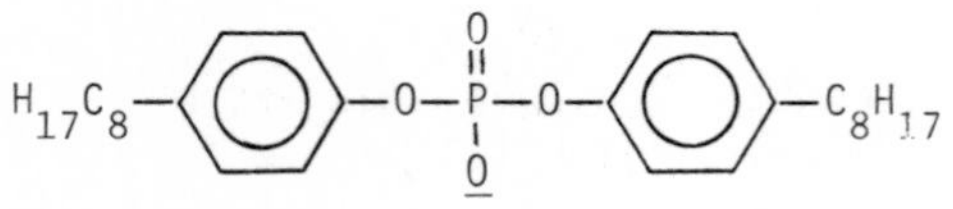

which is selective for M^{2+} over M^+. Dioctyphenylphosphate exhibits good selectivity for Ca^{2+} over Mg^{2+}, K^+ and Na^+ (but not over Zn^{2+}, Pb^{2+}, and Fe^{2+}). The organic phosphate diester is an ion-carrier which provides a mechanism of transport for the Ca^{2+} through the organic phase. Ion exchange equilibria of the type

$$M^{2+}(aq) + CaR_2(oct) \rightleftharpoons MR_2(oct) + Ca^{2+}(aq)$$
$$2H^+(aq) + CaR_2(oct) \rightleftharpoons 2RH(oct) + Ca^{2+}(aq) \qquad [4]$$

must lie far to the left in order for the electrode to exhibit a Nernstian response to $Ca^{2+}(aq)$. Furthermore the exchange rate between $Ca^{2+}(aq)$ and $CaR_2(oct)$ must be rapid in order for the electrode to exhibit reasonable response times.

The calcium ion-selective electrode exhibits a Nernstian slope (29.6 mV per unit in log $a_{Ca^{2+}}$ at 25°C) over the range 1 to 10^{-5} in $a_{Ca^{2+}}$. Between 10^{-5} and 10^{-6} M $Ca^{2+}(aq)$, E becomes independent of the concentration of $Ca^{2+}(aq)$. The lower limit of detection of 10^{-6} M $Ca^{2+}(aq)$ is determined by the solubility of CaR_2 in the aqueous phase. For a fixed concentration of $Ca^{2+}(aq)$ in the external phase, the observed value of E is independent of pH over the range 12 to 5.5. Below pH 5.5 E changes with pH, owing to the transport of H^+ via the complex RH. At low pH values the calcium ion-selective electrode behaves like a pH electrode. The value of k_{Ca^{2+}, Na^+} is about 0.003 for the calcium electrode.

The potassium liquid-ion-exchange electrode, based on the antibiotic valinomycin as a carrier molecule, exhibits a remarkable selectivity for K^+ over Na^+ ($k_{K^+, Na^+} = 1/3800$), because of the stringent space restrictions in the ion cavity (the hole in the center of the molecule). The inside of the valinomycin cavity is ringed with electron rich donor atoms (oxygen and nitrogen), whereas the outer part of the donut-shaped molecule is relatively hydrophobic. The hydrophobic nature of the outer part of the valinomycin-K^+ complex renders the complex soluble in lipid media, such as are found in cell membranes. The antibiotic activity of valinomycin presumably arises from the ability of the molecule to dissolve in cell membranes and collapse the natural K^+ gradient across the membrane. Other antibiotics such as the actins, can also be used to construct ion-selective electrodes. Cyclic polyethers exhibit strong selectivity for ions with diameters comparable to the diameter of the central oxygen cavity. For example, excellent selectivity for Pb^{2+} over Ca^{2+} can be achieved with dicyclohexyl-18-crown-6. A variety of hole sizes can be achieved using other crown ethers and mixed cyclic and polycyclic ethers and amines [10, 11].

Anion selective liquid-ion-exchange electrodes also have been developed [1]. In these cases the ion carrier is a positively charged species, for example, the complex ion

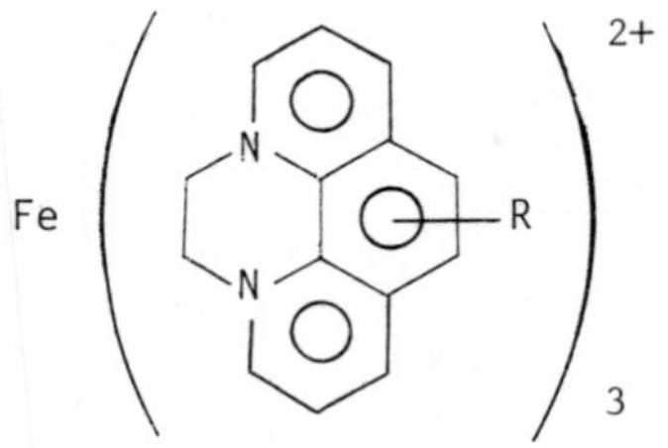

is used in a ClO_4^- ion-selective electrode. The $k_{ClO_4^-, X^-}$ values for several X^- species are: OH^- (1.0), I^- (1.2×10^{-3}), NO_3^- (1.5×10^{-3}), and Cl^- (2.2×10^{-4}).

II. Gas Sensing Electrodes

The first gas sensing electrode system was developed by Severinghaus. This electrode (actually an electrochemical cell) can measure either the CO_2 gas pressure directly in the gas phase, or the concentration of CO_2 in a solution. The concentration of CO_2 is determined through a pH measurement. The CO_2 sensor is shown in cross section below

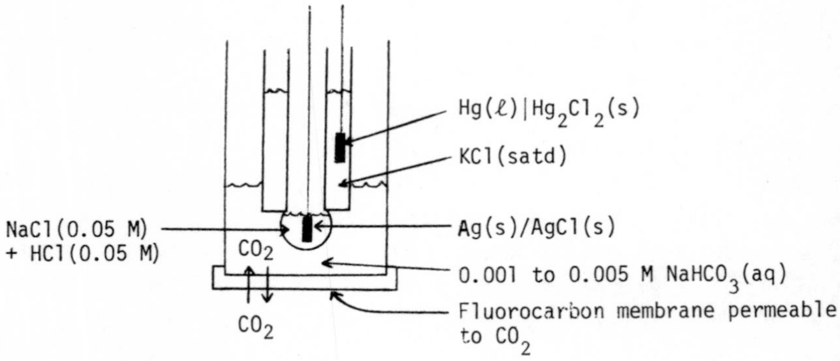

Fig. 2. Cross-sectional view of a CO_2 g sensor.

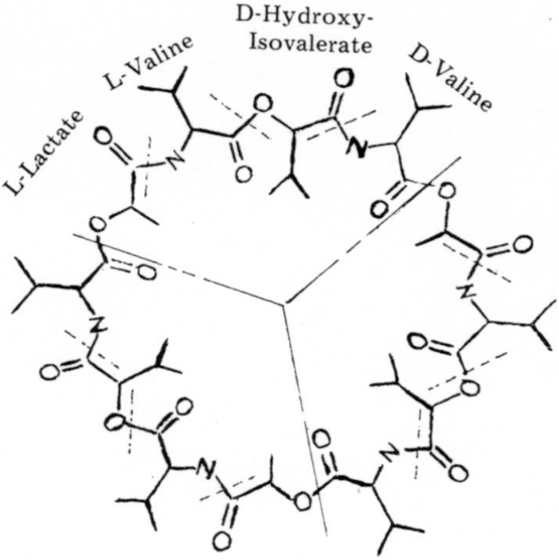

valinomycin

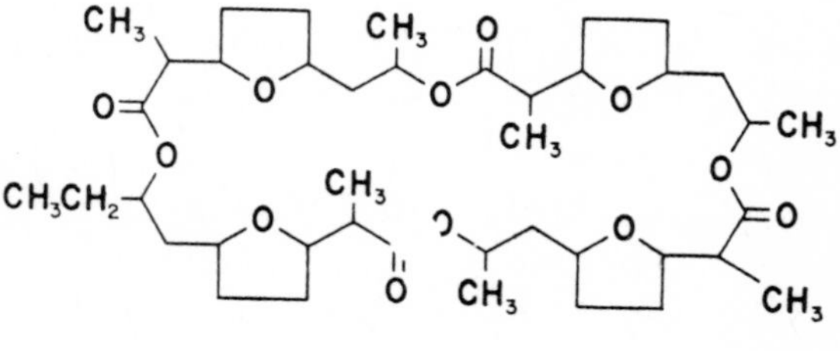

monactin

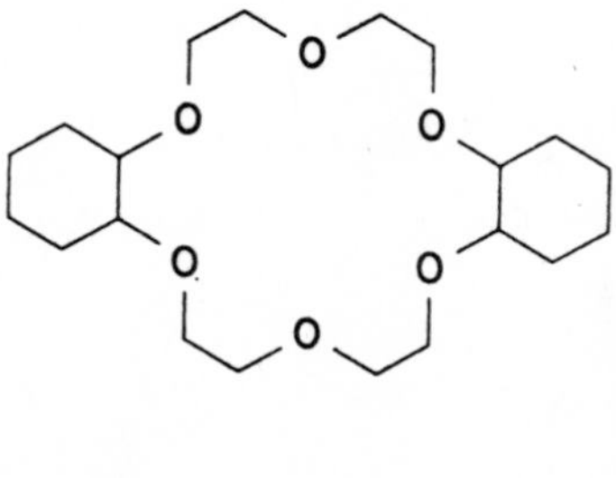

dicyclohexyl-18-crown-6

Carbon dioxide influences the acid-base equilibria in the solution via t
reaction:

$$CO_2(aq) + OH^-(aq) \rightleftharpoons HCO_3^-(aq)$$

for which

$$K = \frac{(HCO_3^-)}{(CO_2)(OH^-)}$$

and $[K_w = (H^+)(OH^-)]$

$$(CO\ ) = \frac{(HCO_3^-)}{K(OH^-)} = \frac{(HCO_3^-)(H^+)}{K\ \ K_w}$$

Because the amount of CO_2 that passes through the membrane is small, (HC
does not change appreciably and $(CO_2) \simeq K'(H^+)$. Thus a measurement of t
pH of the internal $NaHCO_3$(aq) solution, which is in equilibrium across t
fluorocarbon membrane with an external CO_2 source, is equivalent to a
measurement of $-\log(CO_2)$. The principal interferants are small neutral
acidic or basic molecules, such as SO_2, NH_3, and NO_2, which also influen
the acid-base equilibria in the $NaHCO_3$ solution. Systems analogous to th
CO_2 sensor can be set up to measure ammonia and sulfur dioxide concentrations by using dilute NH_4Cl(aq) and $NaHSO_3$(aq) solutions, respectively,
place of $NaHCO_3$(aq).

Because oxygen is neither acidic or basic, oxygen cannot be determi
by coupling to an acid-base equilibrium. Furthermore, oxygen electrodes
are reversible only under conditions of extremely pure electrode solutio
In the Beckman Oxygen Analyzer the concentration of oxygen is determined
polarographically. A cross section of the polarographic cell is shown below.

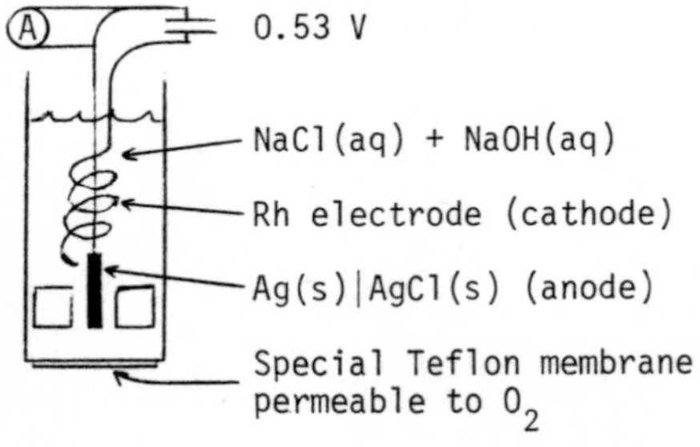

Fig. 3. Cross-sectional view of a polarographic O_2 analyzer.

constant potential of 0.53 V is applied across a rhodium cathode and a ilver/silver chloride anode. The electrode reactions are

$$e^- + O_2(aq) + 2H_2O(l) \longrightarrow 4OH^-(aq) \quad \text{(cathode)}$$

$$Ag(s) + 4Cl^-(aq) \longrightarrow 4AgCl(s) + 4e^- \quad \text{(anode)}$$

t constant applied potential the current flow is proportional to the con-entration of dissolved O_2. The proper functioning of the system depends n the establishment of a steady state for oxygen flow through the membrane, nd, consequently, in solution-phase measurements it is essential to stir he solution. The principal interferants are species that can penetrate the embrane and be reduced at 0.53 V; namely, NO_x, SO_2, and NO_2^- (NO_2^- gives ise to NO_x).

II. Crystal Membrane Systems

Lanthanum fluoride, LaF_3, conducts fluoride ions at room temperature; ne fluoride conductance can be increased by doping the crystal with Eu^{2+}. ne fluoride conductivity of LaF_3 is the basis of the Orion Research fluor-de ion-specific electrode, a cross-section of which is shown below:

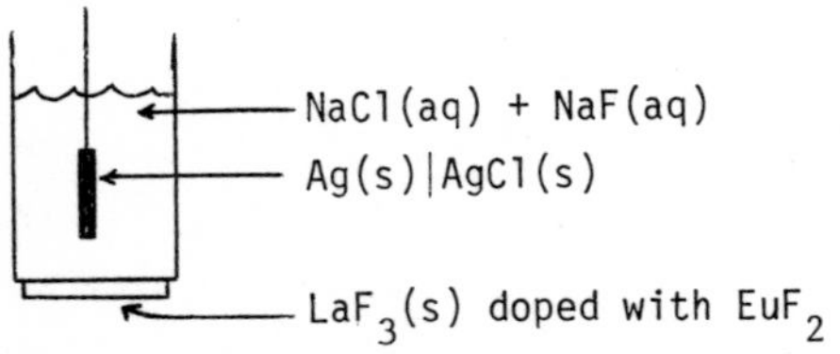

ig. 4. Cross-sectional view of fluoride-specific electrode.

ne electrode processes are

$$g(s) + Cl^-(int) \longrightarrow AgCl(s) + e^-$$

$$F^-(LaF_3) \longrightarrow F^-(int)$$

$$F^-(ext) \longrightarrow F^-(LaF_3)$$

d the net electrode reaction is

$$(s) + Cl^-(int) + F^-(ext) \longrightarrow AgCl(s) + F^-(int) + e^-$$

For a particular electrode all of these quantities are constant at a particular temperature and pressure except F^-(ext) and therefore

$$E = E^\circ + \frac{RT}{F}\ln a_{F^-} \quad [5]$$

The fluoride-ion-specific electrode exhibits a Nernstian response to fluoride ion over the range 1 to 10^{-6} M. The major interferants are OH^- and species that form strong complexes with La^{3+}, such as citrate ion. Hydroxide ion interference presumably arises in two ways: (1) OH^- is similar in size to F^- and OH^- may be capable of moving through the LaF_3 crystal; and (2) at sufficiently low values of the ratio $(F^-)/(OH^-)$ hydroxide can convert LaF_3(s) to $La(OH)_3$(s) via the reaction

$$LaF_3(s) + 3OH^-(aq) = La(OH)_3(s) + 3F^-(aq)$$

$$K = \frac{K_{sp(LaF_3)}}{K_{sp[La(OH)_3]}} = \frac{(F^-)}{(OH^-)^3}$$

where the parentheses denote equilibrium values. For a particular solution of interest

$$Q = [F^-]^3/[OH^-]^3$$

If $Q/K < 1$, then OH^-(aq) can convert LaF_3 to $La(OH)_3$, whereas if $Q/K > 1$, then the LaF_3 is thermodynamically stable in the solution. In practice the condition $(OH^-) < 0.1\ (F^-)$ must be met in order to obtain satisfactory performance.

In the absence of F^- in the external solution the fluoride ion-specific electrode can be used as a lanthanum electrode.

$$3Ag(s) + 3Cl^-(int) \longrightarrow 3AgCl(s) + 3e^-$$

$$3F^-(LaF_3) \longrightarrow 3F^-(int)$$

$$LaF_3(s) \longrightarrow La^{3+}(LaF_3) + 3F^-(LaF_3)$$

$$La^{3+}(LaF_3) \longrightarrow La^{3+}(ext)$$

or

$$3Ag(s) + 3Cl^-(int) + LaF_3(s) \longrightarrow 3AgCl(s) + 3F^-(int) + La^{3+}(ext) + 3e^-$$

a particular electrode at fixed temperature and pressure all of these ies have constant activity except La^{3+}(ext) and

$$E^\circ - \frac{RT}{3F} \ell n\, a_{La^{3+}}$$

Citrate ion is a potential interferant with the fluoride electrode [1] the reaction

$$(s) + Cit^{3-}(ext) \;=\; LaCit(ext) + 3F^-(ext)$$

which

$$K_{sp[LaF_3]} \cdot K_{comp} \;=\; \frac{(F^-)^3(LaCit)}{(Cit^{3-})}$$

stability condition of the LaF_3 membrane is

$$\frac{]^3_{ext}[LaCit]_{ext}}{t^{3-}]_{ext}K} > 1$$

this condition is not met, then Cit^{3-} will produce F^- in the external ution until equilibrium is attained.

Crystal membrane silver and sulfide ion-specific electrodes are based the ionic conductivity of Ag_2S crystals in which Ag^+ ions are the mobile cies. Silver sulfide has a very low solubility in water ($K_{sp} = 10^{-51}$ 25°C) and Ag_2S can be pressed into water-impervious pellets. The range response of the electrode is from saturated solutions down to 10^{-8} M. the Ag^+ or S^{2-} is involved in complexation equilibria in solution, then e ion concentrations can be determined down to about 10^{-20} M. The in-nal electrode element involves a silver wire attached to the Ag_2S pellet. internal electrolyte solution is $Na_2S(aq)$. When electrode functions as ilver electrode, the electrode reactions are:

$$s) \longrightarrow Ag^+(Ag_2S) + e^-$$
$$(Ag_2S) \longrightarrow Ag^+(ext)$$

$$s) \longrightarrow Ag^+(ext) + e^-$$

reas when functioning as a sulfide electrode, the electrode reactions :

$$2Ag(s) \longrightarrow 2Ag^+(Ag_2S) + 2e^-$$
$$S^{2-}(ext) + 2Ag^+(Ag_2S) \longrightarrow Ag_2S(s)$$
$$2Ag(s) + S^{2-}(ext) \longrightarrow Ag_2S(s) + 2e^-$$

The Ag_2S pellet isolates the internal Ag(s) electrode from the component of the external solution, and thereby prevents the interaction with exte nal oxidizing and reducing agents.

The ionic conductivity of Ag_2S can be utilized to construct a varie of anion-selective electrodes in which Ag_2S + AgX mixtures are used as m branes (in some cases mixed salts may form in the pellet, such as Ag_3SX, but mixed-salt formation is not necessary for the functioning of the mem brane). The K_{sp} of the salt imbedded in the Ag_2S matrix must meet the c dition $K_{sp(Ag_2S)} \ll K^2_{sp[AgX]}$, because otherwise X^-(ext) will convert Ag to AgX via the reaction

$$Ag_2S(s) + 2X^-(ext) = 2AgX(s) + S^{2-}(ext)$$

for which

$$K = \frac{K_{sp[Ag_2S]}}{K^2_{sp[AgX]}} = \frac{(S^{2-})}{(X^-)^2}$$

X^-(ext) will react with Ag_2S in the membrane until $(S^{2-})_{ext}/(X^-)^2_{ext}$ K reaches unity. If K is very small, then $(S^{2-})_{ext}$ will be very small. T AgX salt need not be a Ag^+ ion conductor, however, the K_{sp} of AgX must b small enough that the solubility of AgX does not interfere with the dete mination of X^-(ext). In effect $Ag_2S \cdot AgX$ membranes behave as X^- conduct through Ag^+. The electrode reactions are:

$$Ag(s) \longrightarrow Ag^+(Ag_2S) + e^-$$
$$X^-(ext) + Ag^+(Ag_2S) \longrightarrow AgX(s)$$

or

$$Ag(s) + X^-(ext) \longrightarrow AgX(s) + e^-$$

Some examples of $AgX \cdot Ag_2S$ systems are: $AgCl \cdot Ag_2S$ for Cl^-(ext); $AgBr \cdot A$

Br^-(ext); $AgI \cdot Ag_2S$ for I^-(ext); and $AgSCN \cdot Ag_2S$ for SCN^-(ext). Many her such systems appear possible, for example, $AgOOCR \cdot Ag_2S$. The major erferants are those anions Y^- in the external solution that can convert (s) to AgY(s) under the prevailing $[X^-]/[Y^-]$ ratio, or species that form ong complexes with Ag^+. In the former case

$$X(s) + Y^-(ext) \;=\; AgY(s) + X^-(ext)$$

$$= \frac{K_{sp[AgX]}}{K_{sp[AgY]}} = \frac{(X^-)}{(Y^-)}$$

the values of $[X^-]_{ext}$ and $[Y^-]_{ext}$ are such that $[X^-]_{ext}/[Y^-]_{ext} < K$, then e above reaction is spontaneous from left to right and the electrode is operable. Thus for the $AgBr \cdot Ag_2S$ membrane some of the major interferants e I^-, S^{2-}, $NH_3[Ag(NH_3)_2^+$ formation], and $CN^-[Ag(CN)_2^-$ formation].

Cation-selective electrodes based on the Ag^+ conductivity of Ag_2S can so be constructed. The membrane components are MS(s) + Ag_2S(s) for an $^+$(ext) electrode. The electrode reactions are

$$g(s) \longrightarrow 2Ag^+(Ag_2S) + 2e^-$$

$$(s) + 2Ag^+(Ag_2S) \longrightarrow Ag_2S(s) + M^{2+}(ext)$$

$$g(s) + MS(s) \longrightarrow Ag_2S(s) + M^{2+}(ext) + 2e^-$$

r which

$$= \; E° - \frac{RT}{2F} \ell n \, a_{M^{2+}}$$

om this reaction it is evident that $MS \cdot Ag_2S$ membrane electrodes are electrochemically equivalent to electrodes of the third kind [12]. Because of e possible reaction of M^{2+}(ext) with Ag_2S(s) in the membrane

$$_2S(s) + M^{2+}(ext) \;=\; MS(s) + 2Ag^+(ext)$$

e K_{sp} of MS must meet the condition

$$K = \frac{K_{sp[Ag_2S]}}{K_{sp[MS]}} \ll 1$$

Some examples of electrode systems of this type are $CuS \cdot Ag_2S$, $PbS \cdot Ag_2S$, and $CdS \cdot Ag_2S$. Major interferants are cations, N^{2+}, that form sulfides more insoluble than MS(s)

$$N^{2+}(ext) + MS(s) = NS(s) + M^{2+}(ext)$$

$$K = \frac{K_{sp[MS]}}{K_{sp[NS]}} = \frac{(M^{2+})}{(N^{2+})}$$

Special cases of interference can arise in which both cations and anions from the external solution react with the membrane to produce an i soluble phase or soluble complex, for example, the $MS \cdot Ag_2S$ membranes can be destroyed by solutions containing $M^{2+}(aq)$ and a halide, $X^-(aq)$, via th reaction

$$Ag_2S(s) + M^{2+}(aq) + 2X^-(aq) = 2AgX(s) + MS(s)$$

for which

$$K = \frac{K_{sp[Ag_2S]}}{K^2_{sp[AgX]}\ K_{sp[MS]}} = \frac{1}{(M^{2+})(X^-)^2}$$

The stability condition for the membrane is

$$\frac{1}{[M^{2+}][X^-]^2 K} > 1$$

Crystal membrane ion-selective electrodes usually exhibit remarkable selectivities because the mobile ion must occupy lattice sites with sharp defined size and charge requirements.

IV. Heterogeneous (Salt-in-Rubber-Matrix) Systems

The first non-glass ion-selective membranes developed for use in ion selective electrodes were reported by Pungor [13]. A Pungor-type membran consists of a heterogeneous mixture of an insoluble salt containing the i of interest and a silicone rubber. The most satisfactory electrode per-nance is usually obtained with 50 to 60 wt. % salt. The salt-rubber ture can be cured in the barrel of a plastic syringe from which the tip has been remobed. The desired membrane disc thickness can then be ex- ded with the syringe plunger and cut off with a sharp knife. The disc then attached with "rubber cement" to the end of a glass tube. The in- nal electrolyte and electrode are inserted into the tube, and the elec- de lead is held in place with rubber cement. A specific example is the assium-selective electrode based on $K_2Zn_3[Fe(CN)_6]_2 \cdot 8H_2O$ in silicone ber [14]. The K^+ selectivity of this electrode is comparable to the ionic glass electrode. The electrode is easy to prepare and relatively xpensive. Because operational Pungor-type electrodes can be prepared h salts that do not necessarily exhibit appreciable ionic conductivities, membrane ion-transport mechanism may involve the movement of ions along stal surfaces. The relatively high proportion of salt to rubber may be onsequence of a requirement of crystal-crystal contacts in the membrane. all insoluble salts of the ion of interest are equally effective. For mple $KB(C_6H_5)_4$ in silicone rubber is not selective for K^+ over Na^+. gor has shown that heterogeneous membranes composed of a ion-carrier plex such as KCL•valinomycin in a rubber matrix exhibit selectivities parable to the analogous liquid-ion-exchanger membranes.

In general the ion-selectivities of heterogeneous salt-rubber mem- nes are not as good as that exhibited by single crystal membranes, vever, in many cases suitable ionically conducting single crystals are known, and in such cases heterogeneous membrane electrodes are the best ilable systems.

. Glass Membrane Systems

The first operational ion-selective membrane electrode was the hydrogen- -selective glass electrode [15]. The hydrogen glass electrode consists a special composition, thin-walled, glass bulb filled with an aqueous ution containing NaCl and HCl. A silver-silver chloride electrode dips to the internal solution. The glass membrane on contact with aqueous so- tion becomes wet to a depth of about 1000 Å on the inside and outside rfaces. Between the two wet layers, the glass is dry. Ion-exchange uilibria are set up between monovalent cations and protons in the in- rnal solution and the internal wet layer, and the external solution and e external wet layer, e.g.

$$(aq, ext) + M^+(glass, wet\ ext) \rightleftharpoons H^+(glass, wet\ ext) + M^+(aq, ext).$$

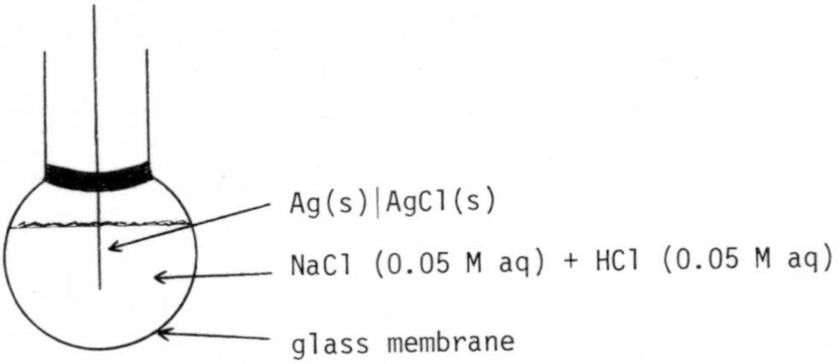

Fig. 5. Cross-sectional view of hydrogen ion glass electrode.

By the use of radioactive tracers it has been established [16] that the current through the dry glass is carried primarily by Na^+. Current throu the wet glass, presumably is carried primarily by H^+. The establishment the external wet layer requires about an hour or so of soaking in an appr priate conditioning electrolyte. The wet layers spall off more or less continuously but slowly while in contact with solution. A glass electrod bulb should never be wiped (it should be cleaned by rinsing), because wip severely disrupts the external wet layer.

A sequence of electrode processes that is consistent with the experi mental data for the hydrogen glass electrode is

$$Ag(s) + Cl^-(int, aq) \longrightarrow AgCl(s) + e^-$$

$$H^+(int, aq) \longrightarrow H^+(glass, inside)$$

$$Na^+(glass, inside) \longrightarrow Na^+(glass, outside)$$

$$H^+(glass, outside) \longrightarrow H^+(ext, aq)$$

The sum of the above four reactions yields

$$\begin{aligned} Ag(s) + Cl^-(int, aq) + H^+(int, aq) + Na^+(glass, inside) \\ + H^+(glass, outside) \longrightarrow AgCl(s) + H^+(glass, inside) \\ + Na^+(glass, outside) + H^+(ext, aq) + e^-. \end{aligned}$$

long as the proton sites in the wet layers are essentially saturated h hydrogen ions, i.e., provided M^+(aq, ext), and OH^-(aq, ext) are not high, then all of the activities in the above reaction are constant ept for H^+(ext, aq), for a given glass electrode at a fixed temperature pressure. Under these conditions the hydrogen glass electrode

s)|AgCl(s)|NaCl(aq), HCl(aq)|H(glass)|H^+(ext, aq)

be represented formally as

lass)|H^+(ext, aq)

lass) ⟶ H^+(ext, aq) + e^-

which

$$= E^\circ - \frac{RT}{F} \ell n\, a_{H^+(ext,\ aq)}.$$

The ion-selectivities of glass electrodes can be varied to a certain ent by varying the composition of the glass [16]. Variations in glass position may give rise to variations in ion site size and geometry. The roximate glass compositions of the three principal types of glass electrodes, expressed as percentage of oxides, are

TYPE	Na_2O	Al_2O_3	SiO_2
lass)	32	< 1	68
(glass)	11	18	71
glass) or cationic glass)	27	5	68

value of k_{Na^+,K^+} for the Na(glass) electrode is about 0.001, and the ue of k_{K^+,Na^+} for the K(glass) electrode is about 0.1.

I. Multilayered, Species-Selective Membranes Involving Enzymes

At present the most active area of research in the field of species-ecific electrodes involves the use of enzymes to measure electrochemically ecies of biological interest ("bioprobes" or enzyme electrodes) [2,3].

The basic idea involved in enzyme electrodes is to combine the speci-city property of a particular enzyme with the selectivity of one of the various types of species-specific electrode systems described in Sections through V above. The enzyme is immobilized in a thin gel layer that is separated from the solution containing the substrate by means of a cellophane membrane. The enzyme-gel layer is in contact with an underlying crystal membrane, or fluorocarbon membrane, or acetate filter disc (Figure 6).

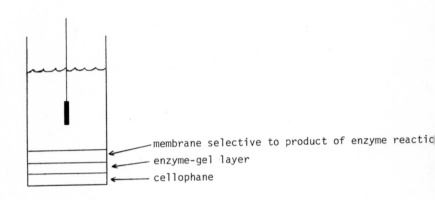

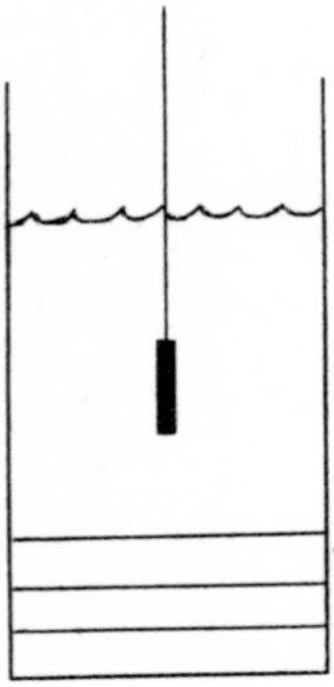

Fig. 6. Multilayered enzyme electrode.

In general, the electrode is designed to monitor (either potentiometricall or polarographically) a <u>product</u> of the enzyme reaction.

Polarographic enzyme electrodes have been used to measure glucose, uric acid, ethanol, and cholesterol. The glucose electrode, which is avai able commercially, employs the enzyme glucose oxidase. The hydrogen peroxide reaction product is determined polarographically by measuring the anodic current arising from the oxidation of H_2O_2

$$C_6H_{12}O_6 \xrightarrow{\text{glucose oxidase}} H_2O_2 + \text{products}$$

$$H_2O_2 \longrightarrow O_2 + 2H^+(aq) + 2e^-$$

Numerous oxygen oxidoreductase enzymes produce H_2O_2 which can be determin polarographically (see the discussion of the oxygen electrode in Section I For example, ℓ-phenylalinine can be determined using ℓ-aminoacid oxidase ich also produces H_2O_2 from the enzyme-catalyzed reaction of ℓ-phenyla-ne with oxygen. In either case the H_2O_2 produced is allowed to pass rough a fluorocarbon membrane into the polarographic chamber where the ectrode potential is fixed at a voltage appropriate for the oxidation of drogen peroxide.

Urea can be determined with a potentiometric enzyme electrode con-ining the enzyme urease

$$\underset{\underset{O}{\|}}{NC}NH_2(soln) + 2H_2O + H^+(aq) \xrightarrow{\text{urease}} HCO_3^-(aq) + 2NH_4^+(aq)$$

this case there are several possibilities for the determination of the ount of product produced. A cationic glass, or valinomycin liquid ion-changer, electrode can be used to determine the concentration of NH_4^+ (K^+ a major interferant); or, an ammonia gas electrode can be used to measure e concentration of NH_3 (extensive calibration would seem to be required cause the ammonia gas electrode also responds to CO_2).

It is possible to combine two enzyme systems in the same gel layer and ereby convert a product of one enzyme reaction into a more conveniently terminable species using a second enzyme reaction

$$1 \xrightarrow{E_1} S_2 \xrightarrow{E_2} P$$

or example, E_1 might be an esterase and E_2 an oxidase.

It has recently been shown [2] that the crystal membranes of some ion-elective electrodes can be used to determine protein concentrations. For xample, the potential of a Ag_2S membrane electrode is shifted when proteins ontaining disulfide groups are present in the solution. Presumably the roteins absorb on the surface of the Ag_2S membrane and alter the electro-hemical potential of the Ag^+ ions in the Ag_2S by forming S-Ag-S type inkages between the crystal and the protein [2] (variation of the Pallman ffect). The shift in electrode potential is linear in the protein con-entration, at least at low concentrations of protein, although saturation ffects undoubtedly occur at higher concentrations.

Enzyme electrodes possess several significant advantages over more onventional analytical methods. The principal advantages are small sample ize (as small as a drop or less) and direct measurement on physiological luids. The technology of miniaturization has advanced to the point where easurements on single cells, at least the larger ones, are now possible [17].

Enzyme electrodes should, of course, be used only with thorough consideration of the possible sources of error. The lower limit of detection is fixed by the sensitivity of the underlying electrode species sensor; th upper limit of detection is usually fixed by substrate saturation of the enzyme. Response time of enzyme electrodes are generally of the order of minutes. Although many enzymes exhibit remarkable selectivities, perfect selectivity is unattainable and possible interferants (enzyme inhibitors o competitive substrates) in the sample should be carefully considered.

The developments and discoveries in the field of species-species elec trochemical sensors over the past decade have been truly remarkable. The prospects for further advances and novel applications are excellent.

REFERENCES

1. R.A. Durst (Ed.), Ion-Selective Electrodes, NBS Special Publication #314 (U.S. Department of Commerce, U.S. Government Printing Office) 1969.
2. G. A. Rechnitz, C&EN, Jan. 27, 1975, p. 29.
3. R. L. Rawls, C&EN, Jan. 5, 1976, p. 19.
4. J. Koryta, Ion-Selective Electrodes, Cambridge University Press (1975
5. Analytical Methods Guide, Orion Research Inc., 380 Putnam Ave., Cambridge, MA (issued periodically, 7th Edition May 1975).
6. IUPAC Bulletin No. 43, January 1975. IUPAC Secretariat, Bank Court Chambers 2/3 Pound Way, Cowley Centre, Oxford OX4 3YF, U.K.
7. M. J. Brand and G. A. Rechnitz, Anal. Chem. 42 (1970) 616.
8. P. A. Rock, J. Chem. Educ. 52 (1975) 787; 47 (1970) 683.
9. R. G. Bates, J. Pure and Applied Chem. 36 (1974) 407.
10. C. J. Pedersen, J. Amer. Chem. Soc. 89 (1967) 7017.
11. R. M. Izatt, J. H. Rytting, D. P. Nelson, B. L. Haymore, and J. J. Christensen, Science 164 (1969) 443; J. Amer. Chem. Soc. 93 (1971) 1619.
12. G. Kortüm, Treatise on Electrochemistry, Elsevier, Amsterdam (1965).
13. E. Pungor, Anal. Chem. 39 (1967) 29A.
14. P. A. Rock, T. L. Eyrich, and S. Styer, J. Electrochem. Soc. 124 (197 530.
15. D. A. MacInnes and M. Dole, Ind. Eng. Chem. Anal. Ed. 1 (1929) 57.
16. G. Eisenman (Ed.), Glass Electrodes for Hydrogen and Other Cations, (Marcel Dekker, Inc. New York, 1967).
17. J. L. Walker, Anal. Chem. 43 (1971) 89A.

MECHANISMS OF ELECTROCHEMICAL OSCILLATIONS

JOEL KEIZER
Chemistry Department, University of California, Davis, CA 95616

ABSTRACT

Spontaneous oscillations of current under potentiostatic conditions or of the voltage under galvanostatic conditions have been widely observed in electrochemical systems. Related oscillations occur in corroding systems that exhibit active-passive behavior and in systems involving shape changes such as the beating mercury heart. Establishing a mechanism for an oscillation in terms of molecular processes requires: (1) Determination of all the variables which are changing. (2) Determination of the kinetic processes which are occurring. (3) Determination of the differential equations which describe how the variables are coupled by the kinetic processes. (4) Analysis and solution of the differential equations and comparison with experimental oscillations. A description of the use of graphical techniques for carrying out the mathematical analysis in Step 4 is given. The use of these ideas is discussed for proposed oscillation mechanisms involving the electrocapillary effect, a negative resistance cell coupled to an impedance, and single electrode processes.

I. INTRODUCTION

Oscillations have been observed in a wide variety of electrochemical systems [1]. For example, with a fixed current of 20μA a voltage oscillation of amplitude 1.2 V and a period of about 5 min. is found in the reduction of formate ion in aqueous KOH at palladium [2]. Similarly the current through a hanging mercury drop electrode in aqueous In^{+3} and SCN^- undergoes oscillations of roughly 7 mA with a 2 sec. period when the external voltage is sufficiently negative [3]. Electrochemical oscillations can also occur in the absence of external circuitry. A particularly captivating example of this is the "beating mercury heart" [4], so called because of the rhythmic motion of the mercury electrode which accompanies the oscillations in this system.* Another example of this sort is the active-passive behavior of an iron wire corroding in nitric acid [6].

The diversity of systems which exhibit electrochemical oscillations well as the diverse properties of the oscillations is bewildering [1]. Indeed only a few electrochemical oscillations can be said to be understo in a mechanistic sense, and it seems clear that these mechanisms have no immediate application to other types of electrochemical oscillations. Co sequently a general, unified description of electrochemical oscillations cannot be given here. Instead a more modest goal will be undertaken, namely, describing what is involved in establishing the mechanism of an electrochemical oscillation. Although the task is really no different fr establishing a mechanism for a chemical reaction, oscillations are usuall associated with the simultaneous occurrence of a number of kinetic process and it is necessary to understand each kinetic process individually befor the resultant oscillation can be understood.

II. ELUCIDATING A MECHANISM

To establish a mechanism for an electrochemical oscillation involves both experimental and theoretical work. The purpose of the experimental work is to determine what is happening in the system. In particular one must first determine (i) what variables are changing. These variables may include voltages, current, concentrations, capacitances, surface coverages or any number of observables that can be measured experimentally. Once it is clear what variables are changing, it is necessary to establish (ii) th molecular processes which cause these variables to change. Since electrochemical oscillations can be caused by all sorts of processes, the method of determining which processes are occuring is really up to the wit of the experimentalist, and this task may be formidable. In fact, this is the key step in elucidating an oscillation mechanism, and a good knowledge of the electrochemical processes which occur in simple systems is the only guide for success here.

---

*The mercury heart was observed first by Kühne [5] and is easily constructed. Pour a 3 mm diameter drop of mercury into a watch glass and cover th mercury with 2 M H_2SO_4(aq.) to which a few crystals of $K_2Cr_2O_7$ have been added. Rest a clean, sharp iron nail on the side of the watch glass under the solution, and adjust the point of the nail until it just touches the mercury. After minor adjustment of the position of the nail, the oscillations will occur.

Although good answers to the experimental question (i) and (ii) would m to establish an oscillation mechanism, they do so only in a tentative se. What remains to be verified is that the proposed mechanism actually roduces the experimental results. This requires (iii) formulating the etic-differential equations which correspond to the mechanism and (iv) ving these equations and comparing to experiment. For an oscillating tem these steps are complicated and quite different from the usual kin-c analysis of simple systems. For example, if a chemical reaction mech-sm has an experimentally measured first order rate law, one has already lved" the differential equations by making the usual semi-log plot of data. On the otherhand, electrochemical oscillations involve coupled inary or partial differential equations whose mathematical analysis is from straight-forward. This "theoretical" part of establishing a mech-sm is vital, and a proposed mechanism which does not have oscillating utions to its differential equations is certainly incorrect.

These steps in the elucidation of an oscillation mechanism are ustrated by the work of Lin, et al. [4] on the mercury heart oscillation.

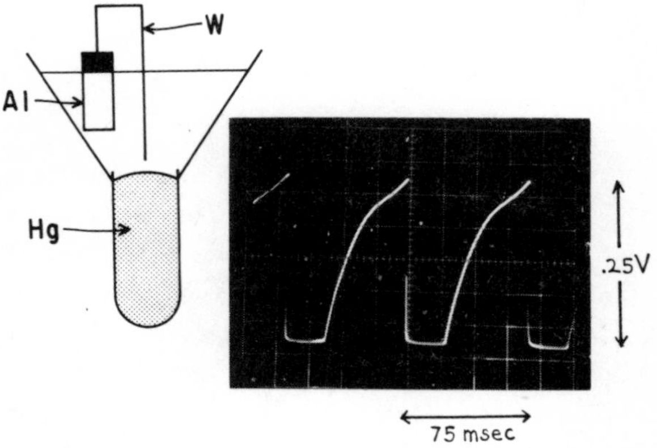

g. 1. The modified beating mercury heart system studied by Lin, et al.]. The voltage oscillations were recorded on an oscilloscope.

The system they investigated is shown in Fig. 1, and corresponds to the ce diagram $A\ell|W|OH^-(aq)$, $O_2(aq)|Hg(\ell)$ with a voltage $\phi_{Hg} - \phi_{W(A\ell)} = 1.27$ V. When the sharpened tungsten electrode is brought into contact with the column of mercury, the surface of the mercury begins to move and the voltage oscillations shown in Fig. 1 are observed.

By measuring the voltage of the mercury and tungsten separately with respect to a reference electrode it was discovered that only the voltage the mercury changed appreciably during the oscillations [4]. It was also discovered that during the oscillations the pH of the solution above the mercury increased. The source of the OH^- ions was found to be the mercur surface, and it was concluded that another important variable is the surface concentration of OH^- on mercury. The shape change of the mercury meniscus was observed under magnified, slow-motion photography [7], and t shape of the meniscus was found roughly to remain part of a hemisphere. Consequently the height of the center of the meniscus is a third importan variable. The only other variable which changes significantly is the concentration of O_2 dissolved in the solution.

The key kinetic process involved in the oscillations is the reducti of oxygen on mercury. This deduction was based on the observation that a solution purged with an inert gas did not oscillate and that the amplitud of the oscillations is a linear function of the oxygen concentration. Th reduction of oxygen is also compatible with the increase in pH through th net reaction

$$O_2 + 2H_2O + 4e^- = 4OH^-.$$

Since the constant voltage portion of the oscillation (see Fig. 1) corresponded to $\phi_{Hg} - \phi_{W(A\ell)} \approx 0$ and since the slow-motion photography showed th the tungsten actually touches the mercury, it was determined that the met lic conduction of electrons (short circuit) occured. Finally since surfa concentrations and voltage effect the surface tension of mercury (the "ele trocapillary effect" [8]), it was deduced that the mercury surface was moving because of a changing surface tension. Based on the slow-motion photography results the inertia of the mercury column was felt to be unimportant [7].

These observations led to the following "mechanism" for the oscillations [4]: (a) Production of electrons at the corroding aluminum through the half-reaction

$\ell(s) + 4OH^-(aq) = A\ell(OH)_4^-(aq) + 3e^-$

) Transfer of these electrons to the mercury via metallic contact with e tungsten. (c) Reduction of O_2 on mercury. (d) Desorption of the OH^- duction product from the mercury, and (e) Mercury shape changes driven by changing surface tension. The significant variables changing in this chanism are v = the Hg - $A\ell$(W) voltage difference, s = the Hg-W tip sep- ation, and c = the surface concentration of OH^- on mercury.

This mechanism was compatible with the observations as far as they nt but required a theoretical analysis in order to show that it gave both alitative and quantitative agreement with experiment. The kinetic equa- on correspond to this mechanism are [4]

$$/dt = L\big(\gamma(c,v) - \gamma(s)\big) \quad [1]$$

$$/dt = [i_\circ \exp(-v/v_\circ) - i'_\circ \exp(-v/v'_\circ)] - D(c-c_\circ) \quad [2]$$

$$v/dt = [i_\circ \exp(-v/v_\circ) - i'_\circ \exp(-v/v'_\circ)] - \sigma(s)v. \quad [3]$$

the first equation L is a proportionality constant relating the rate of ange in separation to the difference between the actual surface tension, c,v), and the equilibrium surface tension at that height, $\gamma(s)$. The cond equation represents the production of OH^- ions on mercury by the duction of oxygen as well as their removal by desorption. Finally, Eq. [3] lates the change in voltage to capacitance (C) and short circuit effects, th $\sigma(s)$ the conductance between the mercury and tungsten tip.

These three kinetic equations are coupled and because they are non- near they cannot be solved analytically. Some of their properties, wever, are easy to see. For example, at the moment of short circuit the nductance σ almost instantaneously reaches some large value σ_{max}. This ans that the metallic conductance term dominates the reduction term in . (3) and leads to

$$/dt = -\sigma_{max} v/C.$$

us the short-circuiting should be a first order process with a time con- ant $\tau_{s.c} = C/\sigma_{max} \approx 10^{-7}$ sec for this system. This is compatible with e measured time scale of the rapid decrease to short circuit in Fig. 1.

Actually a more detailed analysis of these equations is possible [7]. Because of the rapidity of the short circuit, an excellent approximation to the solutions of these equations has been obtained. The equations show a limit cycle oscillation, that is, a completely periodic solution for certain initial values of c, v, and s. A comparison of this solution with the experimentally measured oscillations is given in Fig. 2. The quality of the agreement with the experimental oscillations suggests that the proposed mechanism is correct.

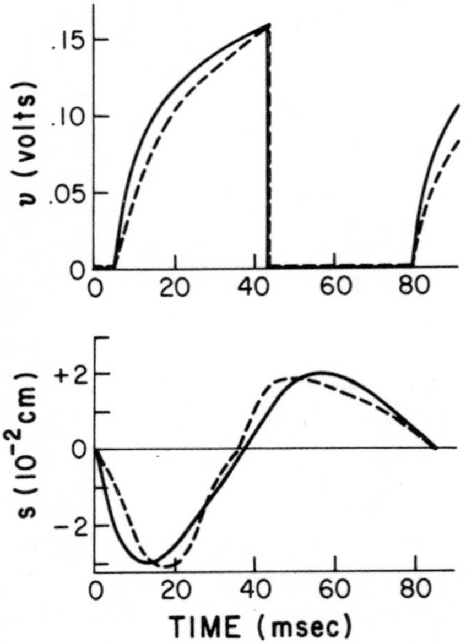

Fig. 2. Voltage and separation oscillations for the mercury heart [7]. The dashed lines are from experimental measurements and the full lines were calculated for the limit cycle of Eqs. [1] - [3].

III. MATHEMATICAL ANALYSIS OF OSCILLATIONS

Although a great many experimental facts are known about electrochemical oscillations, only a few proposed mechanisms have been given careful mathematical analysis. As the analysis of the mercury heart oscillation ws, this is an essential part of verifying a mechanism and the remainder this paper will be concerned with some simple ways of analyzing coupled etic equations.

One of the most useful techniques for treating coupled differential ations is a graphical phase space analysis [6]. This analysis depends y on the form of the differential equations and is best explained with example. Consider the "two-variable" version of the mercury heart model in Eqs. [1] - [3]. In this version of the model it is assumed that the aration between the tungsten and the mercury relaxes very rapidly so that dt = 0 <u>always</u>. Consequently Eq. [1] implies

$$) \quad = \quad \gamma(c,v). \tag{4}$$

s eliminates s as an independent variable since the functional relation- p in Eq. [4] can be solved to give $s = f(c,v)$. Using the further sim- fication in Eq. [2] that $c_\circ = 0$ and ignoring the anodic half reaction on cury gives the "two-variable" model equations:

$$\text{'dt} \quad = \quad i_\circ \exp(-v/v_\circ) - Dc \tag{5}$$

$$\text{/dt} \quad = \quad i_\circ \exp(-v/v_\circ) - \sigma\big(f(c,v)\big)v. \tag{6}$$

solutions to these equations are functions of time c(t) and v(t) which pend on the values of c and v at t = 0. This can be represented graph- ally by a curve in a two dimensional c-v "phase plane", as illustrated Fig. 3. The parameter which generates this curve is the time, whose lues are not explicitly given by this representation. In this way the lution to Eqs. [5] and [6] can be represented by a path or trajectory in e phase plane. In particular a trajectory which is a closed loop repre- nts a periodic solution and so an oscillation [10].

A point on the path in the phase plane is given by a position vector $t) = \big(v(t),\, c(t)\big)$, and the rate of change of $\vec{P}$ with time is $d\vec{P}/dt = \vec{V} =$ v/dt, dc/dt). Because Eqs. [5] and [6] are differential equations they ve explicit expressions for the velocity $\vec{V}$ at any point (v,c), namely,

$$v,c) \quad = \quad (C^{-1}[i_\circ e^{-v/v_\circ} - \sigma\big(f(c,v)\big)v],\; i_\circ e^{-v/v_\circ} - Dc). \tag{7}$$

The idea of the phase plane analysis is to draw the velocity vectors a variety of points in the phase plane to get an idea of the shape of

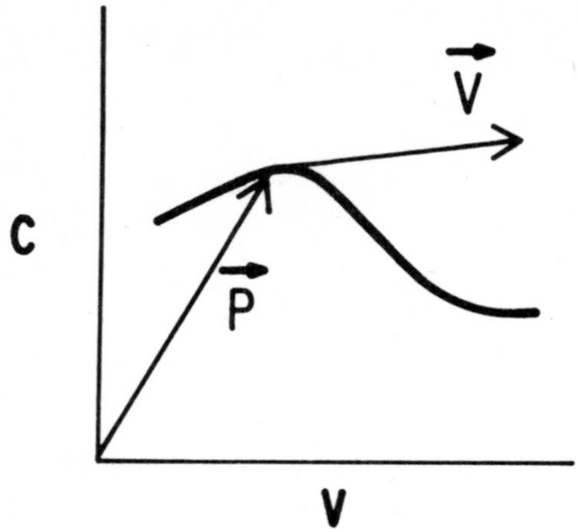

Fig. 3. The phase plane for the "two-variable" mercury heart model. The tip of the vector $\vec{P}$ locates a point of the solution path of Eqs. [5] and [6]. $\vec{V}$ is the velocity (dv/dt, dc/dt).

the trajectories [10]. This is analogous to using dye streak lines in a moving fluid to help visualize the flow. Actually a more complete pictur is gotten by concentrating on certain points or lines in the phase plane. For example, the isoclines are the curves on which either dv/dt = 0 or dc/dt = 0, i.e., where the velocity vectors $\vec{V}$ all point in either the c-direction or the v-direction [10]. Eq. [7] shows that the equations of the isoclines are given by

$$dv/dt = 0: \quad \sigma\big(f(c,v)\big)v = i_\circ \exp(-v/v_\circ) \qquad [8$$

$$dc/dt = 0: \quad c = i_\circ \exp(-v/v_\circ)/D. \qquad [9$$

The points of intersection of the isoclines are the steady states, so-call because at these points both dv/dt = 0 and dc/dt = 0. Thus at the steady states the velocity $\vec{V}$ is zero and v and c are independent of time. A stea state can be stable or unstable depending on whether the phase plane poin near a steady state stay close or move away from it. This is reflected in velocity vectors $\vec{V}$ that point toward or away from the steady state.

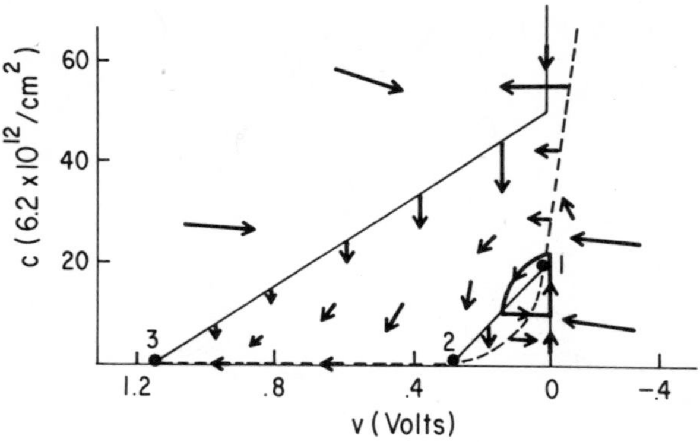

Fig. 4. Graphical analysis in the phase plane of the "two-variable" mercury heart model. The closed trajectory around the steady state 1 is the limit cycle.

This graphical analysis has been carried out in Fig. 4 for certain parameter values of the "two-variable" model. The isocline dc/dt = 0 is given by the dashed line and the full line is the dv/dt = 0 isocline. The isoclines intersect at the points 1, 2, and 3 giving three steady states. The velocity vectors converging on steady state 3 show it to be stable and the ones diverging from steady states 1 and 2 show that they are unstable. Another important feature visible in Fig. 4 is the "swirling motion" around steady state 1. In fact, a detailed calculation [7] shows that a unique closed trajectory exists in this region. This trajectory is a limit cycle, and it is stable in the sense that points near, but off, the limit cycle move onto it as time proceeds. This trajectory corresponds to a periodic oscillation of the variables c, v, and s -- a complete calculation giving the results in Fig. 5.

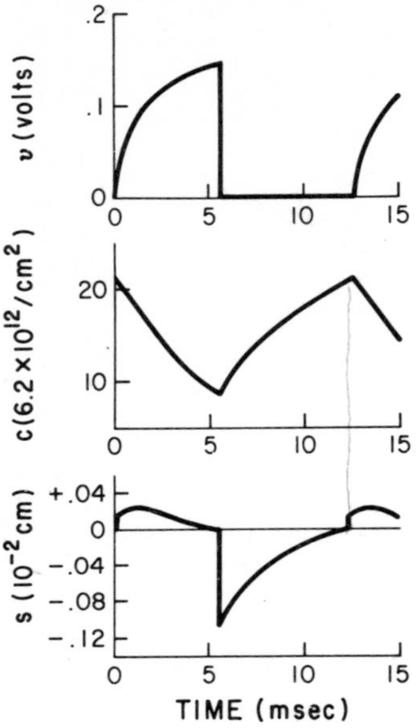

Fig. 5. Oscillations for the "two-variable" mercury heart model based on the limit cycle in Fig. 4 [9].

It was just this sort of graphical analysis on the full three-variabl model in Eq. [1] - [3] that gave the first indication that the proposed mechanism for the mercury heart would produce oscillations. The graphica analysis for those equations is somewhat more complicated, but proceeds without diffiuclty on a blackboard with colored chalk.

IV. EXTERNAL CIRCUIT OSCILLATIONS

Electrochemical oscillations can occur without the direct influence the electrochemical kinetic equations. The mathematical analysis of such oscillations then depends only on the circuit equations and is rather simple. An oscillation of this sort occurs for a cell composed of a pping Hg electrode in an aqueous solution of 1 mM $In(NO_3)_3$ and 1 M KSCN h an Hg pool counterelectrode [11]. When the external circuit maintains ixed voltage E between an induction coil and a resistor (see Fig. 6) usoidal oscillations of the cell voltage V are observed. The period of oscillations is the order of 5 msec and the amplitude is roughly 8 mV.

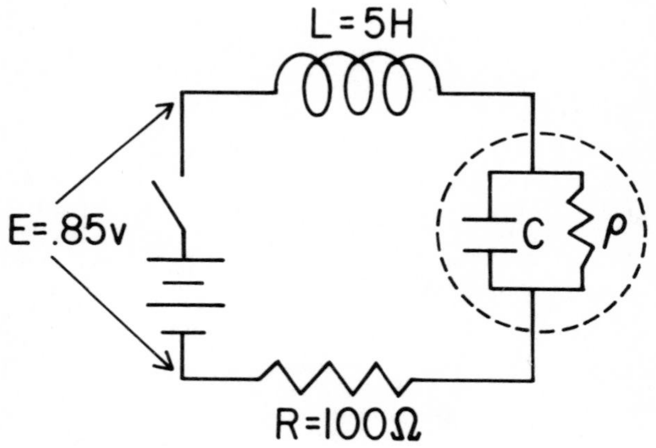

. 6. Equivalent electrical circuit for the In^{+3}, SCN^- oscillator [11,
.

In this system only the cell voltage V and cell current i vary apprebly and the proposed mechanism for the oscillations involves the negative istance characteristic of this electrochemical cell [3,11]. The rate ations which correspond to this mechanism follow from Kirchoff's law lied to the equivalent circuit in Fig. 6, namely,

$$= L di/dt + Ri + V$$

$$= i_C + i_R$$

$$= V/\rho$$

$$i_C = C\,dV/dt$$

where ρ is the cell resistance which is <u>negative</u>, C is the capacitance o the dropping Hg electrode, V is the cell voltage, and L is the inductanc These equations can be combined to give the coupled kinetic equations fo V and i:

$$dV/dt = i/C - V/\rho C \qquad [$$

$$di/dt = (E-V)/L - Ri/L. \qquad [$$

Neglecting the voltage dependence of ρ, these equations are easily analy since the right hand sides are then linear functions of i and V. The unique steady state in this system is found by setting $dV/dt = 0$ and $di/dt = 0$ and solving the resulting simultaneous equations to obtain $V^{ss} = E\rho/(R+\rho)$ and $i^{ss} = E/(R+\rho)$.

The phase plane analysis is actually simpler when the new variables $\delta V = V - V^{ss}$ and $\delta i = i - i^{ss}$ are introduced. This substitution into Eqs. [10] and [11] gives

$$d\delta v/dt = -\frac{1}{\rho C}\,\delta v + \frac{1}{C}\,\delta i \qquad [$$

$$d\delta i/dt = -\frac{1}{L}\,\delta v - \frac{R}{L}\,\delta i. \qquad [$$

The two dimensional phase plane of this system is the $(\delta v, \delta i)$ plane and steady state is at the origin $\delta v = \delta i = 0$. The $d\delta v/dt = 0$ isocline is gotten from Eq. [12] as the straight line $\delta i = \delta v/\rho$, and the $d\delta i/dt$ isocline is $\delta i = -\delta v/R$. Since ρ is negative and R is positive, the phase plane graphs in Fig. 7 are obtained. Two cases are shown there. The gr for which $R/|\rho|>1$ clearly exhibits a swirling motion around the steady state, whereas the graph $R/|\rho|<1$ shows that the steady state is unstable

Actually Eqs. [12] and [13] can be solved in closed form since they are linear. The solution is a linear combination of the exponentials $e^{-\omega_1 t}$ and $e^{-\omega_2 t}$ where ω_1 and ω_2 are eigenvalues of the "rate constant matr

$$\begin{pmatrix} -\frac{1}{\rho C} & \frac{1}{C} \\ -\frac{1}{L} & -\frac{R}{L} \end{pmatrix}$$

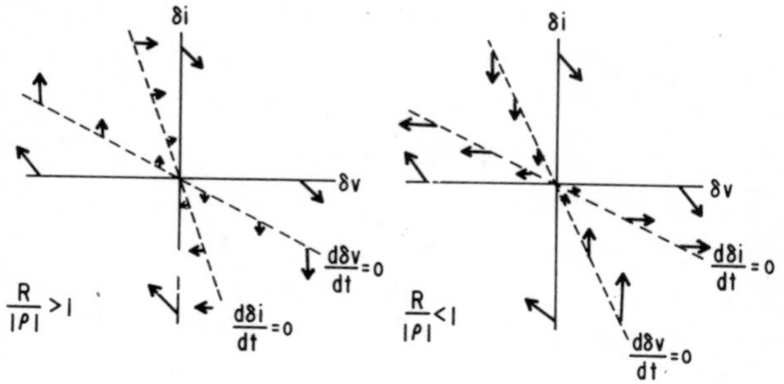

ig. 7. Phase plane analysis for the In^{+3}, SCN^{-} oscillator for $R/|\rho|>1$ and $/|\rho|<1$.

When ω_1 or ω_2 are complex with a negative real part, the steady state unstable and the solution has a sinusoidal factor. This will occur when th

$$\left(\frac{1}{\rho C} - \frac{R}{L}\right)^2 < \frac{4}{LC}, \quad \frac{1}{\rho C} + \frac{R}{L} < 0 \qquad [14]$$

ich is the criterion (in the linear approximation) for the attainment of cillations in this system [11]. Of course, under these conditions the tual motion in the phase plane is not unbounded as the linear analysis ggests. A complete description of the trajectories requires a knowledge the dependence of the cell resistance ρ on the voltage. Nonetheless, e frequency calculated by the linear analysis as well as the conditions quired for oscillation in Eq. [14] are in good agreement with experiment 1].

V. OSCILLATIONS INVOLVING SINGLE ELECTRODE KINETICS

The mechanism for the mercury heart oscillations does not require ci cuit analysis because no external circuitry is necessary for the oscillations. On the other hand, the In^{+3}, SCN^{-} electrode oscillations described in the previous section required only an analysis of the circuit equations and is independent of the underlying electrochemical mechanism. In fact, this sort of analysis can <u>always</u> be carried out if one is willin to treat an electrochemical cell as equivalent to an electronic circuit element with frequency dependent response functions [3]. In this sense a electrochemical oscillation can be explained by circuit analysis using experimentally determined admittances. This, however, is not in the spirit of the explanation outlined in Section II of this paper and does not get the molecular basis of the oscillations. In fact what is hidden in the frequency dependent admittances are time dependent molecular process whos details constitute the mechanism.

The really complicated -- and, thus, interesting -- electrochemical oscillations undoubtedly involve only single electrode processes. An oscillation of this sort was discovered by De Levie [3] using a modification of the In^{+3}, SCN^{-} electrode discussed above. The modified electrod is made by increasing the SCN^{-} concentration to 5 M, increasing the $In(NO_3)_3$ concentration to 1.2 mM and adjusting the pH to 3.6. Removing t induction coil gives the circuit diagram

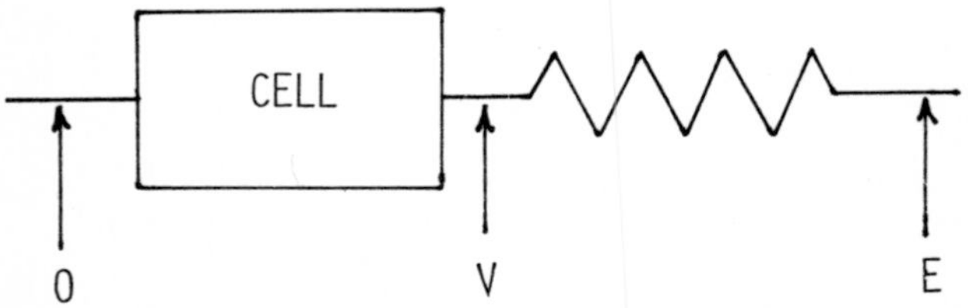

where the voltage E is held fixed. This device shows current oscillation of about 7 μA with a 2-5 sec. period when E is fixed in the range -.99 V -.71 V with respect to a Ag/AgSCN counter electrode. Since the induction coil has been removed from the circuit, the only possible mechanism for t oscillations involves the kinetic process occuring at the electrode.

Unfortunately the analysis of a real oscillator whose mechansim involves only electrochemical processes at a single electrode has not been carried out [1]. Consequently to illustrate the nature of the problems involved a hypothetical mechanism will be introduced. Although the model may be applicable to a real system, it should be regarded as purely pedagogical. The electrode processes are supposed to mimic passivation and the electrode reaction is taken to be

$$X_{(sol)} + nA_{(ads)} \rightleftharpoons XA_n^{+m} + me^-.$$

The species XA_n^{+m} is supposed to form a surface film, $A_{(ads)}$ is an adsorbed species or an electrode atom, and X is present in the solution. Diffusion

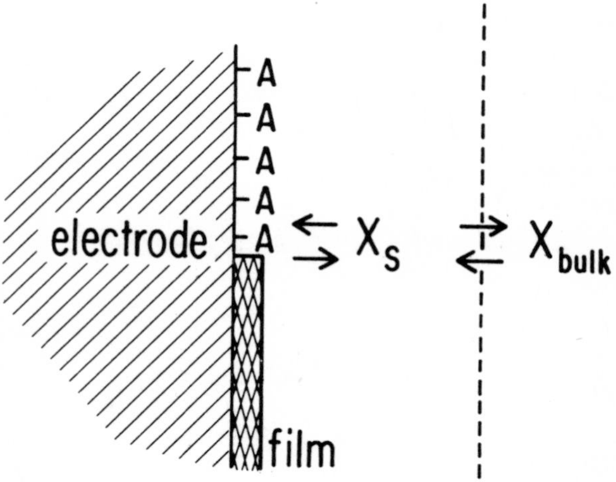

ig. 8. Pictoral representation of the diffusion layer of length ℓ for the ectrochemical passivation process.

fects on X are described by a "diffusion layer" of length ℓ as indicated Fig. 8. The cell circuit diagram is

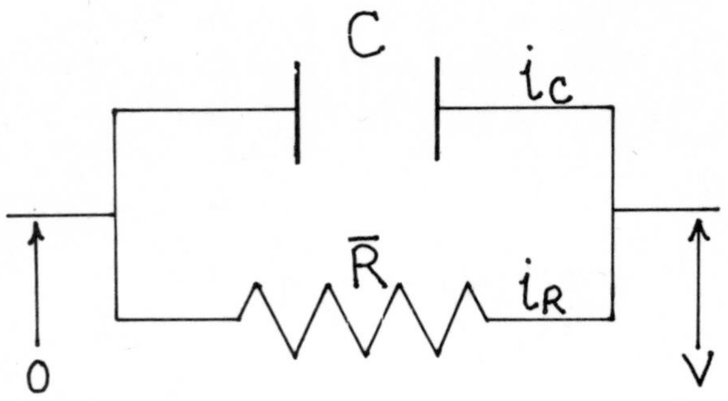

where the capacitive current of the electrode is i_c = CdV/dt and the resistive current is

$$i_R \;=\; k'(1-\Theta)\exp(-v/v_o') - kX_s\Theta\exp(v/v_o). \qquad [1$$

This is the Butler-Volmer current resulting from the electrochemical reaction and Θ = fraction of the surface covered by A, 1-Θ = fraction of th surface covered with the XA_n^{+m} film, and X_s is the concentration of X in t "diffusion layer" next to the electrode. If this cell is operated potentiostatically the circuit diagram is

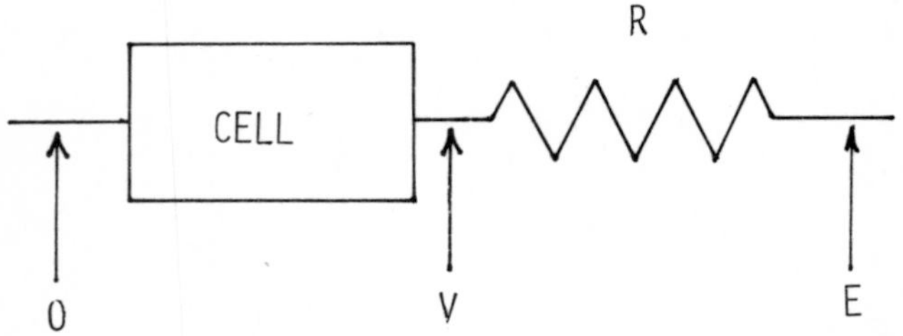

where E is fixed. Using i = i_R + i_C, E = V + iR, and the circuit diagram yields the kinetic equations for this device:

$$dV/dt \;=\; (E - V)/RC - i_R/C \qquad [1$$

$$d\Theta/dt \;=\; Bi_R \qquad [1$$

$$dX_s/dt \;=\; Fi_R - D(X_s - X_{bulk}) \qquad [1$$

where B and F are proportionality constants and D is the diffusion consta of X divided by the length of the diffusion layer ℓ.

Eqs. [16] - [18] couple together the three variables V, Θ, and X_s and are non-linear because of the form of the Butler-Volmer current in Eq. [15]. The reason that these equations are more complicated than Eqs. [12] and [13] is that the cell resistance (see Eq. [15]) depends on the coverage and diffusion layer concentration X_s which are functions of time. This time dependence would cause a frequency dependent admittance in such a system. Although the mathematical analysis of these rate equations has not been undertaken, the results would be interesting since the mechanism includes a variety of single electrode effects.

ACKNOWLEDGEMENT

This work was supported in part by National Science Foundation Grant CHE 74-00483 A03, the Committee on Research of the University of California, Davis, and N.A.T.O. Research Grant No. 696.

REFERENCES

1. A comprehensive review of electrochemical oscillations is found in J. Wojtowicz, in Modern Aspects of Electrochemistry, Vol. 8, J. Bockris and B. Conway (Editors). Plenum Press, New York (1973) p. 47.
2. J. Wojtowicz, N. Marincic, B. Conway, J. Chem. Phys. 48 (1968) 4333.
3. R. de Levie, J. Electroanal. Chem. 25 (1970) 257.
4. S.-W. Lin, J. Keizer, P. A. Rock, and H. Stenschke, Proc. Nat. Acad. Sci. USA 71 (1974) 4477.
5. H. E. Hoff, L. A. Geddes, M. E. Valentinuzzi, and T. Powell, Cardiovas. Res. Bull. (Houston) 9 (1971) 117. A good resumé of the history of the mercury heart is contained in this article.
6. K. F. Bonhoeffer, J. Gen. Physiol. 32 (1948) 69.
7. P. A. Rock, J. Keizer, S.-W. Lin, ms. in preparation.
8. D. Grahame, Chem. Rev. 41 (1947) 441.
9. J. Keizer and J. Schellenbach, unpublished work.
10. N. Minorsky, Non-Linear Oscillations, van Norstrand, New York (1962).
11. R. Tamamushi, J. Electroanal. Chem. 11 (1966) 65.
12. R. Tamamushi and K. Matsuda, J. Electroanal. Chem. 12 (1966) 436.

ELECTROCHEMISTRY OF NERVES*

WALTER J. MOORE

School of Chemistry, University of Sydney N.S.W. 2006, Australia

ABSTRACT

Two kinds of problems are found in the electrochemistry of nerves: (1) steady state membrane potentials, (2) rapid transients associated with the nerve action potential. The steady state regime involves classic electrodiffusion and boundary layer theories, under conditions that seldom permit simplifying assumptions because the thickness of the membrane is comparable with the Debye length in the surrounding solutions. The titration curves of the dissociating groups on the membrane surface depend on the transmembrane potentials. The study of the action potential includes quantities that are strange to physical chemists: conductances that are delayed functions of the electric field strength. Introduced by Hodgkin and Huxley in their phenomenological account of nerve conduction, they have by now achieved the status of basic biophysical concepts. Ultimately these conductances g(E,t) must be explained in terms of the membrane structure. This part of the theory is least developed, but appears to involve the dependence on the electric field of rate constants for conformation changes in macromolecules.

The history of electricity usually begins with static electricity. Amber, the greek <u>electron</u>, rubbed with fur, or glass rubbed with silk gave small shocks and sparks. Electroscopes were devised to show the repulsion and attraction of charged bodies and Leiden jars were developed to store large amounts of static electricity. The principal indication of an electric phenomenon was the visible spark. Only toward the end of the 18th century did the discoveries of Galvani and Volta reveal a new kind of phenomenon, which won the right to be called <u>electric</u> when sparks were achieved from the Voltaic pile. Actually, however, there was a source of galvanic electricity well known to the ancients, the discharge from electric fish. This was described in the scientific literature in 1750 independently by vesande and Adanson. It was not clearly recognized as electricity be-se the key experimental phenomenon, the electric spark, could with dif-ulty or not at all be obtained from the fish.

Electric fish of the family Torpinidae (including Torpedo, Raia, cine), were all common in Mediterranean waters. We can see them depicted the mosaics of Pompeii. Considerably later, fish of the family notidae (including electric eels of the Amazon region) were discovered. of these, Electrophorus electricus, is the prime example of a fresh-er electric fish. The powerful electric discharge of the eel can amount 600 V consisting of about half a dozen pulses each lasting 2 to 3 ms. structure of individual electroplaques and the way in which they gener-high voltage discharges are shown in Fig. 1.

The generation of high voltages across the electric organs of these h is convincing evidence that high electric fields can occur across in-idual cell membranes. The 600 V across the electric organ of a large ctric eel is simply the summation in series of the resting potentials of 90 mV across about 4000 rows of electroplaques. A potential difference 100 mV across a membrane 10 nm thick would correspond to an electric ld of 10^7 V/m. This field is much higher than those usually employed in ctrochemical experiments. It is therefore quite likely that special h field effects may be important in the electrochemistry of nervous tems. Such effects include dipole reorientations through large angles increases in the dissociation constants of weak electrolytes due to the ld (Wien effects).

About 90 years ago Carl Sachs, while studying electric eels in ezuela, noted that when the electric organ became fatigued it still re-nded to external electric impulses imposed upon it. This result was te contrary to the behavior of the electric organ in Torpedo, which was responsive to outside electric stimulation. Thus there is a marked ference between the response of the innervated membrane of Electrophorus that of Torpedo. In Electrophorus the membrane is both chemically and ctrically excitable whereas in Torpedo it is only chemically excitable. both cases the chemical transmitter is acetylcholine. All fresh water ctric fish have membranes that are both chemically and electrically itable, whereas salt water electric fish have membranes that are only mically excitable. Table 1 summarises these differences between classes cell membranes.

Charles Darwin in "Evolution of Species and the Descent of Man" dis-sed the occurrence of electroplaques as one of the so-called "special

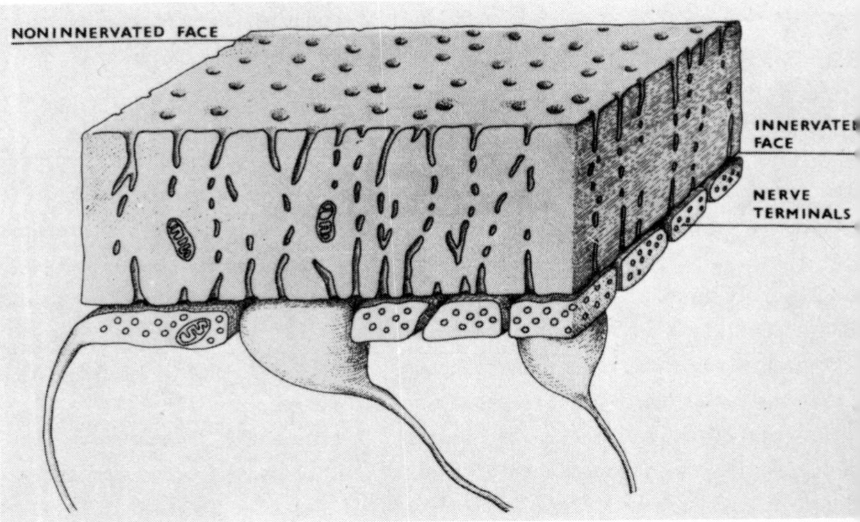

Fig. 1a. Schematic picture of the structure of a section of an electro-plaque cell in electric eel. One face of the cell is densely covered wi nerve terminals. When an impulse comes from the brain of the animal, th terminals release acetylcholine which rapidly diffuses across the gap to the surface of the electroplaque cell, causing the membrane to depolariz

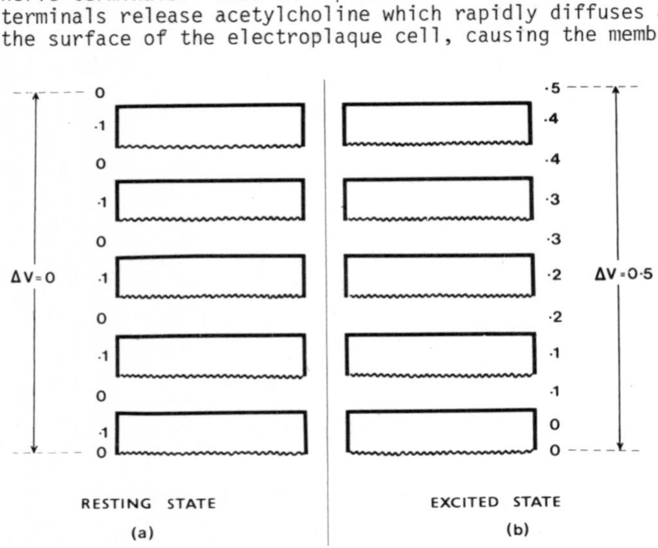

Fig. 1b. Schematic picture of a stack of electroplaque cells. In the resting state the potential differences across opposite faces of each ce cancel, so that the overall p.d. is zero. In the excited state all the innervated faces become depolarized and the p.d.'s across the noninnerva faces are summed to give a large overall p.d. across the electric organ.

Pictures after V. Whittaker in "Cell Membranes". (H.P. Publishing Co., New York, 1975. Eds. G. Weissman and R. Clairborne)

ficulties of the theory of natural selection". While electric organs of eel or Torpedo would be useful for defence, the weak discharge of Raia d be of no such use. Darwin correctly concluded that electric organs lly evolved from muscle cells. The weakly electric fish use electric lses as sensory devices for detecting objects in water. Electric fish special receptors that can detect changes in electric fields as saml 0^{-5} V m^{-1}, far below the threshold of any other known sensory receptors . It is possible that sensory cells for detection of electric fields also exist in land animals, especially birds. This subject is one of at potential interest if indeed these new sensory modalities can be nly established.

The origin of nervous systems and nerve cells must go far back to the ly times of prebiotic evolution. Thus even membranes in non-living tems can display many of the properties of living cells, such as detec- n of external influences, response, conduction and excitation. The liest nerve cells were probably sensory cells which differentiated from thelial cells and allowed the animal to sense specific changes in its ironment and to make appropriate responses. Even protozoa (one called anisms) display many patterns of response to stimulation and it is not d to imagine the specialization of cells to carry out such responses. nitive metazoa (organisms with many cells) developed nerve nets permit- g the signaling of data from one part of the organism to another. Still er these nerve nets condensed into centralized nervous systems.

Table 1

Types of Cell Membranes

itation	Examples
Chemically excitable	1. Innervated membrane of electroplaques of *Raia* and *Torpedo*; slow muscle fibres of amphibians.
Electrically excitable	2. Surface membranes of vertebrate nerve and muscle fibers.
Both chemically and electrically excitable	3. Surface membranes of denervated or embryonic mammalian muscle; surface membrane of crustacean muscle; innervated membrane of electroplaque of *Electrophorus*.
Neither chemically nor electrically excitable	4. Noninnervated membrane of electroplaques of *Electrophorus*.

The Nerve Cell

Nerve cells (also called neurons) come in many different forms and sizes. It is not possible, therefore, to depict a "typical neuron". Figure 2 shows one kind of neuron. This is a motoneuron that transmits impulses to terminals at muscle cells, thus providing the signals that initiate muscular contraction.

The special structures seen in this cell occur in all neurons. The consist of a cell body, an extensive process (which may be branched) cal the axon, and a number of processes called dendrites. The function of t axon is to conduct electrical impulses away from the cell body to neighboring cells. The dendrites provide conducting pathways that bring electrical impulses into the neuron. The cell body integrates the incoming impulses (which may be either excitatory or inhibitory). At a certain critical level of net excitatory input within a given time, the cell "fir and emits a pulse or train of pulses along the axon. The propagated pul is called the nerve action potential. It consists of a wave of membrane depolarization that travels down the axon.

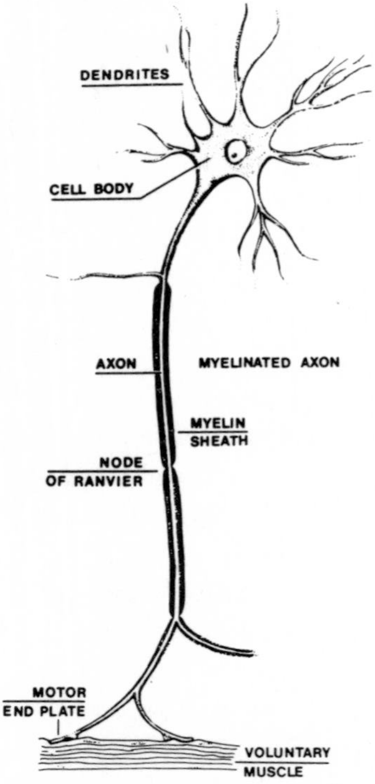

Fig. 2. The essential features of a motoneuron. The length of the myelinated axon would in fact be much longer than shown. This picture is adapted from an illustration by R.J. Demarest in "The Human Nervous System" by C.R. Noback and R.J. Demarest (McGraw Hill, New York, 1975).

The energy required to generate these communication pulses has been ored as electrochemical free energy in the form of gradients of ionic con- ntrations and of electric potential across the neuronal membranes. The terior of the nerve cells is high in K^+ and low in Na^+ and the external dium is high in Na^+ and low in K^+. For example, the intracellular medium cat motoneurons was estimated to have $[Na^+]_i$ = 15 mM, $[K^+]_i$ = 150 mM, and e intercellular fluid, $[Na^+]_o$ = 150 mM, $[K^+]_o$ = 5.5 mM.

When a cell "fires" or a neuronal membrane is "excited", Na^+ ions rush to the cell interior. Permeability of the membrane to K^+ ions also rises, t more slowly. During resting states the ionic gradients are restored, as e Na^+ ions are pumped out of the cell against their gradient of concentra- on ("active transport") by a <u>sodium pump</u>, which derives its free energy m the hydrolysis of ATP by an enzyme that is incorporated into the membrane.

Two kinds of axons are found in both peripheral and central nervous tems. Unmyelinated axons are essentially uncovered (uninsulated) nerve ers, along which action potentials can travel as continuous waves. Mye- ated axons are covered with a tightly wound lamellar sheath of myelin vided by accessory glial cells (which are not themselves neurons). At te regular spacings along these axons the insulating myelin sheaths are errupted by regions of unmyelinated axon called <u>nodes of Ranvier</u>. The tance between the nodes varies from 50 to 1500 μm.

The nerve action potential in myelinated axons can jump from node to e as a result of <u>saltatory conduction</u>. This process is based upon rapid duction down the interior of the axon, with excitation of the axonal brane occurring only at the nodes. Saltatory conduction greatly in- ases the velocity of nerve conduction for a fiber of given cross- tional area. To maintain the same conduction velocities in the nerve ers of the human spinal cord without the use of myelinated axons would e required a spinal cord with a diameter comparable to that of a fair ed tree trunk. The use of myelinated axons also greatly decreases the gy requirements of a nervous system, since the influxes of Na^+ ions ch must be pumped out again) occur only at the nodes of Ranvier.

There is good experimental evidence that the excitable membranes of e cells contain at least two kinds of channels, which are highly (but absolutely) selective for the passages of Na^+ or K^+ ions. The Na^+ nel is specifically blocked by certain powerful nerve poisons, particu- y tetrodotoxin (derived from the puffer fish). Quantitative binding ies with radioactive tetrodotoxin have indicated that the surface den- of Na^+ channels is about $5 \times 10^8\ cm^{-2}$ in squid axons and about

4×10^{10} cm^{-2} at nodes of Ranvier of frog nerves. We do not have a high affinity blocking agent for K^+ channels, but $(C_2H_5)_4N^+$ ions are quite effective. The surface density of K^+ channels appears to be abc 10 ti higher than that of Na^+ channels [2].

Transmembrane Potentials

The first electrochemical problem that we meet in the study of livi cells is how to understand the steady-state potentials across the extern cell membranes, the so-called resting potentials, in the absence of any citation. The plasma membranes are typically about 7 nm thick and consi of about 60% lipid and 40% protein. The thickness was chosen by Nature biological rather than mathematical convenience since it falls midway be tween the limits of the thin-film and thick-film approximations in the theory of membrane potentials.

In 1890 Max Planck [3] presented an elegant analysis of the motion through a membrane of ions subject to the combined influences of electri fields and gradients of concentration. The idealized system discussed b Planck is shown in Figure 3. The membrane is modeled by an infinite pla

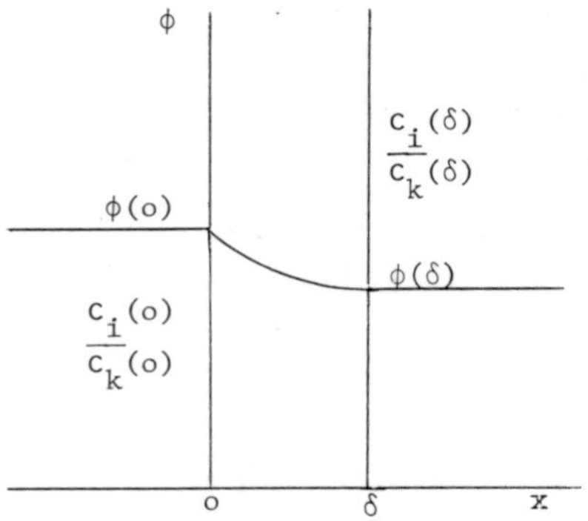

Fig. 3. Conditions for the Planck electrodiffusion problem. The membra extends from $x = 0$ to $x = \delta$, and the concentrations of ions and the elec potentials are constant in the solutions on both sides of the membrane.

region of dielectric constant ε and thickness δ. The electrodiffusion becomes a one-dimensional problem. With rapid stirring, the solutions either side of the membrane are assumed to have uniform concentrations a electric fields right up to the boundary planes at $x = 0,\delta$. The concent tions of the positive ions are designated c_i with $i = 1...n$, and of the negative ions $\bar{c}_k$ with $k = 1...m$.

The steady state is characterized by the condition that in the absence an external electric field there is no net electric current through the mbrane. If I denotes current and J flux, z the charge number, and e the otonic charge, then

$$I_i + I_k) = \sum(z_i eJ_i + z_k eJ_k) = 0 \qquad [1]$$

e flux of each individual ion J_i, J_k is constant in the steady state ney would all vanish only in the equilibrium state). Thus for each ionic ecies Planck wrote

$$= -D_i \frac{dc_i}{dx} + u_i c_i E = \text{constant} \qquad [2]$$

ere D_i is the diffusion coefficient, u_i is the ionic mobility (velocity unit field strength), and E is the electric field.

Planck has thus separated the force acting on an ion into a term de- ding on the gradient of concentration and a term depending on the gra- nt of electric potential. Such a separation involves two approximations: the replacement of a gradient of activity by a gradient of concen- tion (b) the neglect of cross terms that express the dependencies of the x of one component on the gradient of electrochemical potential of other ponents in the solution.

The Nernst equation relates electric mobility to diffusion coefficient

$$= \frac{z_i e}{kT} \qquad [3]$$

s [2] becomes

$$= -D_i \frac{dc_i}{dx} + \frac{D_i z_i e c_i E}{kT} \qquad [4]$$

dividing by D_i

$$\frac{}{i} + \frac{z_i e}{kT} E c_i = A_i \qquad i = 1....n \qquad [5a]$$

re $A_i = J_i/D_i$ is a constant for the steady state of Eq. [2]. Similarly,

$$\frac{\bar{c}_k}{x} + \frac{z_k e}{kT} E \bar{c}_k = B_k \qquad k = 1...m \qquad [5b]$$

ny point in the membrane, the Poisson equation (which relates the di- ence of the electric field to the density of charge) is assumed to hold:

$$\frac{dE}{dx} = \frac{d^2\Phi}{dx^2} = \frac{e}{\varepsilon_o\varepsilon}\sum(z_i c_i + z_k \bar{c}_k) \qquad [6$$

where Φ is the electric potential, ε is the dielectric constant or relativ permittivity, and ε_o is the permittivity of space.

There are m + n flux equations and hence, including Eq. [6], m + n + equations to be solved, with m + n + 1 integration constants. The ionic concentrations at x = 0, δ are all fixed, so that with Eq. [1] we have 2(m + n) + 1 conditions, which will suffice to determine the m + n + 1 integration constants and the m + n ionic fluxes (or constants A_i, B_k). Th problem is thus well defined, but the integration of the set of coupled equations for the general case presents some serious mathematical difficulties.

Planck achieved a major simplification of the problem by restricting his theory to ions having the same charge number $|z|$. Thus he could trea solutions of Na^+, K^+ and Cl^- ions, but not the effect of introducing some Ca^{2+} ions.

Let us take $z_i = 1$, $z_k = -1$ and introduce dimensionless variables

$$\xi = x/\delta \text{ and } \psi = e\Phi/kT$$
$$p = \sum c_i/\bar{C} \text{ and } n = \sum c_k/\bar{C} \qquad [7$$

Here

$$\bar{C} = \langle\sum c_i + \sum \bar{c}_k\rangle = \text{constant}$$

is the space average of the ionic concentrations in the membrane. The Poisson equation [6] now becomes

$$\frac{\bar{\lambda}^2}{\delta^2}\frac{d^2\psi}{d\xi^2} = n - p \qquad [8$$

where

$$\bar{\lambda}^2 = \frac{\varepsilon_o\varepsilon}{e^2\bar{C}} \qquad [9$$

Planck considered the case $\lambda^2/\delta^2 \ll 1$, a condition which becomes mo satisfactory at higher membrane thickness and higher ionic concentration In this case, it would follow from [8] that $p \approx n$, i.e., electric neutra would exist within the membrane even if $d^2\psi/d\xi^2 \neq 0$ (electric field no constant). Goldman (1943) [4] considered the opposite case, $\bar{\lambda}^2/\delta^2 \gg 1$, hich becomes appropriate for thin membranes and lower ionic concentrations. rom [8] the Goldman approximation leads to

$$\psi/d\xi^2 = (n-p)\ \frac{\delta^2}{\bar{\lambda}^2} = 0 \qquad [10]$$

that even if $n \neq p$ (membrane not electrically neutral), the field $E =$ Φ/dx would be constant. Thus the Goldman treatment is often called the nstant-field model.

One might suppose that experimental parameters would at once indicate r choice between Planck and Goldman approximations for nerve membranes. th T = 300 K, $\delta = 7$ and $\varepsilon = 5$ (as appropriate for lipid films),

$$/\delta^2 = \frac{1.462 \times 10^{23}}{\bar{C}}$$

ere $\bar{C}$ is in units of ions per cubic meter.

The difficulty now is that we do not have an experimental value for $\bar{C}$, e average concentration of ions within the membrane. We know that the ectrical conductivity σ of a typical nerve membrane (squid giant axon) is out 7×10^{-8} S m^{-1} in the resting stage. The conductance is determined the concentration of charge carriers $\bar{C}$, their charge q, and their elec- c mobility u, as $\sigma = \bar{C}qu$. An experimental separation of $\bar{C}$ and u for the ve membrane has not been possible and thus we do not know if the membrane tains relatively few charge carriers of high mobility or a larger number carriers of lower mobility.

The resting conductance σ is due mainly to K^+ ions, σ_K. During exci- ion σ_K increases about 20-fold; thus the resting value should not ex- d about 1/20 that in aqueous solution. Hence $u = 3.8 \times 10^{-8}$ m^2 s^{-1} V^{-1} 25°C and the ionic concentration in the membrane would be 5×10^{-6} molar. this would be the concentration averaged over the whole volume of the brane. The channels transfused by the K^+ ions have been estimated at to 6.0×10^{14} m^{-2}, with a diameter of about 5×10^{-10} m. Thus the vol- fraction of the membrane occupied by the K^+ channels would be from 3 12 x 10^{-5}. The molar concentration of K^+ in channels would thus be 1.6 5.0 x 10^{-3} molar. Our best estimate of the ionic concentration in the hannels of the resting membrane would thus be about 5×10^{-3} molar. At concentration the Debye length would be 6.7 nm. Thus $\bar{\lambda}/\delta \simeq 1$ and her the Planck nor the Goldman approximation for electrodiffusion ars to be satisfactory.

Another serious problem is also evident. There seems to be little t in discussing a Debye length in a channel having a diameter much smaller than such a length. In other words, the basic electrodiffusion theory was not designed for motion through channels. The one-dimensional electrodiffusion model may be appropriate for uniform membranes. If a nerve membrane is thought to be a uniform region with a certain average concentration $\bar{C}$ of conducting ions, then the Goldman constant-field model would appear to provide an appropriate electrodiffusion equation, but it can hardly be applied directly to migration of ions through conducting channels in a nonconducting membrane.

The Planck theory was extended by Schlögl [7] to consider ions of di ferent charge types and to include a uniform concentration of fixed charg in the membrane. The Schlögl theory is cumbersome and requires involved numerical computations. It has therefore been little used in neurophysio- logy altho it might have interesting applications in view of the importan physiological effects of Ca^{2+} ions.

Application of Eyring Rate Theory

The diffusion theory of Parlin and Eyring [5] would provide a genera formulation applicable to channels as well as homogeneous membranes. The ions are assumed to jump from one definite site to another and the rate o jumping can be expressed in the formalism of the activated complex theory of reaction rates in terms of a frequency factor kT/h and a free energy o activation $\Delta G^{\neq}$. The frequency of jumping forward to or backward from a single site is

$$k_{\pm} = \frac{kT}{h} \exp\left[\left(-\Delta G^{\neq} \mp \frac{z\lambda F}{2} \frac{d\Phi}{dx}\right)/RT\right] \qquad [1$$

where λ is the distance between sites and F the Faraday constant. In th steady state the fluxes over successive barriers are equal. The interna concentrations at successive sites can be eliminated. Then the flux thr the membrane becomes

$$-J_i = \frac{k_o\lambda\left(C_o - C_n \dfrac{k_1'\cdots k_n'}{k_o\cdots k_{n-1}}\right)}{1 + \dfrac{k_1'}{k_1} + \dfrac{k_1'k_2'}{k_1k_2} + \ldots + \dfrac{k_1'k_2'\ldots k_{n-1}'}{k_1k_2\ldots k_{n-1}}} \qquad [$$

Here C_o and C_n are the ionic concentrations on each side of the mem brane and k_i, k_i' are the forward and reverse rate constants respectively

Many special cases of [12] can be obtained. For example Woodbury [introduced the assumption of constant field into Eq. [11], and evaluate Eq. [12] for a small number of sites separated by suitably chosen barrie

e Goldman Equation

The attractiveness of the constant-field approximation is that it unples the electrodiffusion equations, so that now we have one equation each ion. From [5],

$$- \Phi_o = -E\delta \qquad [13]$$

us consider only univalent ions so that $z_i = +1$ and $z_k = -1$. For any , j,

$$- \eta c_j = -A_j$$

re $\eta = eE/kT$. On integration

$$= \frac{A_j}{\eta} + \text{const. } e^{\eta x}$$

1 the boundary conditions,

$$= 0, \; c_j = c_j(o), \text{ so that } c_j(o) = \frac{A_j}{\eta} + \text{const.}$$

$$= \delta, \; c_j = c_j(\delta) \text{ so that } c_j(\delta) = \frac{A_j}{\eta} + \text{const.} e^{\eta\delta}$$

$$\text{efore, const.} = [c_j(\delta) - c_j(o)]/(e^{\delta\eta} - 1)$$

$$A_j = \eta \left[c_j(o) - \frac{c_j(\delta) - c_j(o)}{e^{\delta\eta} - 1} \right]$$

current,

$$e\,(\sum D_j A_j - \sum D_k B_k)$$

steady-state conditions, I = 0 and

$$\Phi_o = - \frac{kT}{e} \ln \frac{\sum D_j c_j(\delta) + \sum D_k c_k(0)}{\sum D_k c_k(\delta) + \sum D_j c_j(0)} \qquad [14]$$

Altho this equation has a severely restricted range of validity, it een enthusiastically embraced by neurophysiologists. They often fit imental potential differences to the equation by calculating the ratios nic diffusion coefficients that would be required to reproduce the ved $\Delta\Phi$. This procedure can hardly be recommended. Nevertheless the ion may give some guidance as to expected ranges of $\Delta\Phi$ across nerve anes. Suppose a membrane is permeable only to K^+, Na^+ and Cl^- ions and $_{Na^+}$ = 20 and $D_{Na^+} = D_{Cl^-}$. If $C(0) - C(\delta) = 0.2M$, and the ratio of K^+ + ions is 10/1 inside the membrane, at 25°C the calculated $\Delta\Phi = -42.2$ mV.

This value may be compared with the equilibrium potentials calculated fro the Nernst equation. At 25°C

$$V = \frac{-RT}{zF} \ln \frac{c_i}{c_o} = 0.257 \ln \frac{c_i}{c_o}$$

$$V_{K^+} = -68.3 \text{ mV}, V_{Na^+} = +68.3 \text{ mV}, V_{Cl^-} = 0.$$

Goldman-Hodgkin-Katz Equation [8]

In any case where the resting potential does not equal the equilibri potential for an ion, a net transport of that ion will occur across the membrane. In the constant field approximation, the current of each ion would be

$$I_i = e D_i \rho \left(\frac{C_\delta - c_o e^{\eta\delta}}{I - e^{\eta\delta}} \right) \qquad [$$

where $\eta = Ee/kT = e(\Phi_o - \Phi_\delta)/\delta$.

A metal wire can be inserted down the length of an axon, and the potenti across the axonal membrane maintained constant with a feedback device. This arrangement is called a voltage clamp, [9]. The current across the m brane can now be measured following a step in the voltage to a new clamp value. By substituting an impermeant ion such as $Tris^+$ for Na^+, the tota measured current can be divided into its K^+ and Na^+ components on the assumption that the K^+ current is not altered by the substitution for Na

In Fig. 4, Eq. [15] is plotted for K^+ ions across a membrane with $\delta = 7$ nm and $c_o/c_\delta = 20$. Note the marked rectification — the membrane conductance is much greater in the forward direction than in the revers This nonlinearity or nonohmic behavior is not due to any dependence of ionic permeabilities on potential, but rather to the electrodiffusion m and the resulting space charge limitations on the conductances.

The Nerve Action Potential

We shall return to the problems of the steady state distribution o ionic concentrations and the electric field across nerve membranes. Fi we must introduce the most important property of the nerve membrane, up which depends the ability of nerves to transmit information throughout animal organism in the form of pulsed changes in transmembrane potentia

If a depolarizing pulse is applied across a nerve membrane from ar external source of electric current, the potential inside the nerve ri from its resting value of about -70 mV relative to the outside potenti When the depolarizing pulse exceeds a certain critical value of about

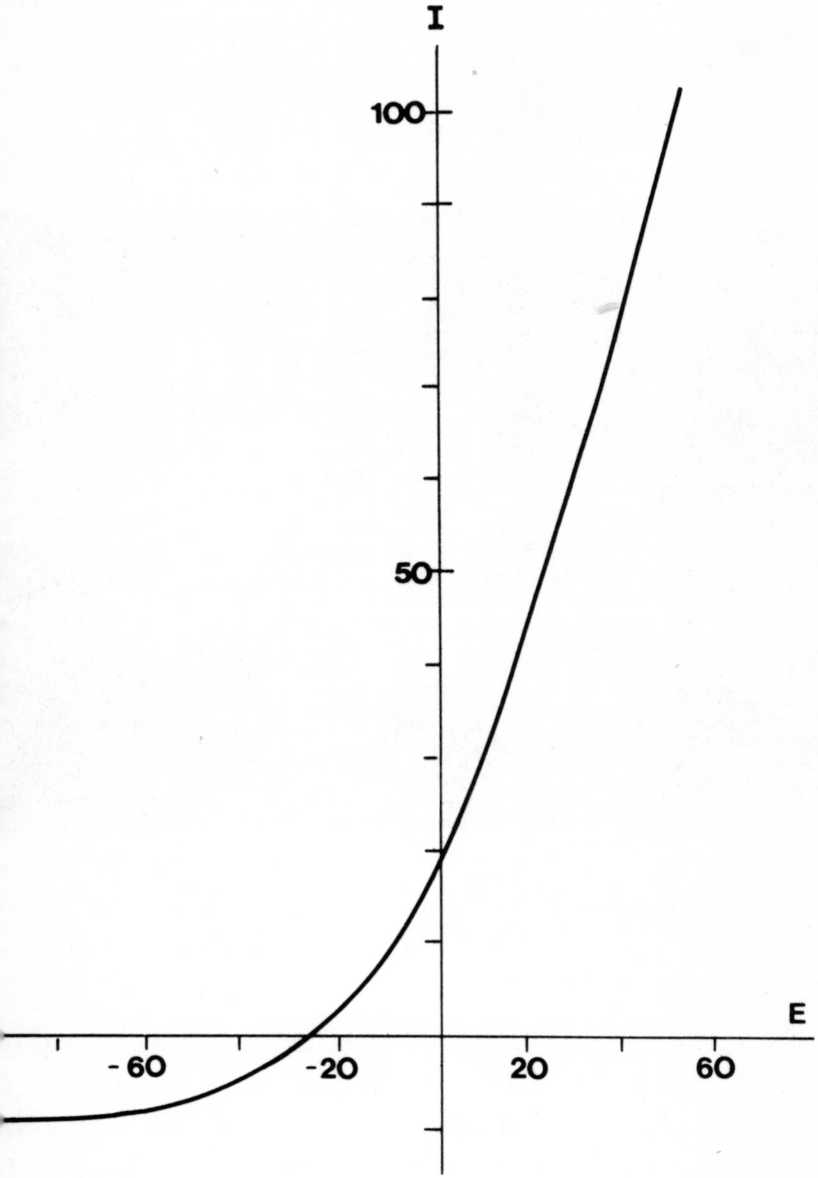

. 4. Current-Voltage characteristic of a membrane according to the dman-Hodgkin-Katz equation [15]. The voltage is in mV and the current arbitrary units.

the nerve membrane may undergo a rapid further depolarization to a maximu[m] value of about +50 mV. This is the basic phenomenon of electrical excita[-]tion of the nerve membrane. It is an all or nothing effect, i.e., the nerve either fires completely or it does not fire at all. There is never [a] "graded response" in which the membrane would fire in an abortive or half hearted way. Some typical results are shown in Fig. 5.

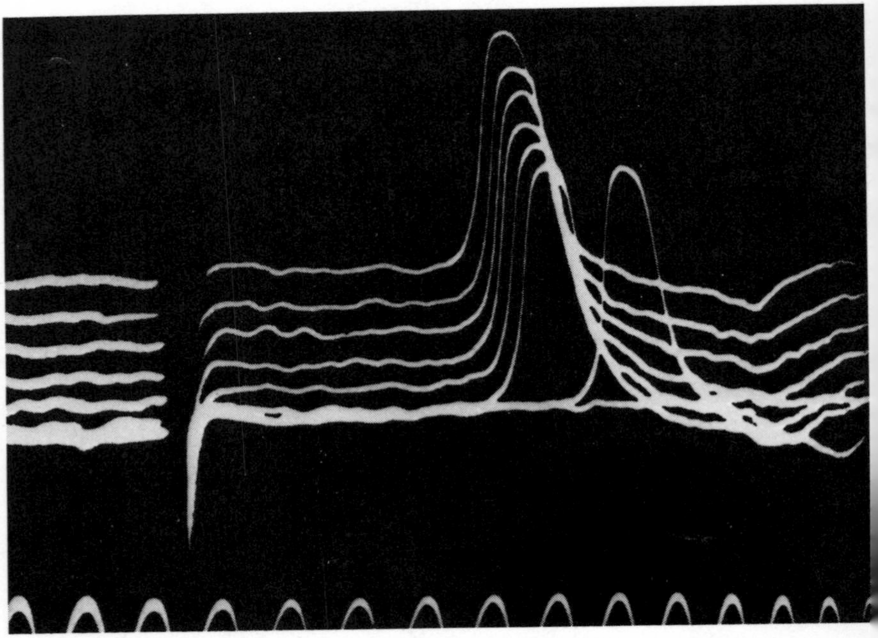

Fig. 5. Experimental recordings of a nerve action potential that illus-trate the "all or none" property of excitation. A single fibre of frog nerve was excited by a current from an external source. From the bottom record upwards the relative current strengths were 1.00, 1.05, 1.50, 2.0[0], 3.00 and 3.50. The base line was shifted upward in successive records i[n] proportion to the strengths of the applied current. The two lowest re-cords (displaced to right) were taken at a current which in one case cau[sed] firing and in the other case caused no response at all. (From. I. Tasak[i], Nerve Excitation, A Macromolecular Approach (Springfield, OH, Charles C Thomas 1968).)

The membrane potential of a squid axon in the resting state is -50 [to] -70 mV, which is close to an equilibrium K^+ potential. The height of th[e] action potential is about +50 mV, which approaches an equilibrium Na^+ po[-]tential. The Goldman equation [14] would explain the changes in membra[ne] potential in terms of changes in the relative permeability of the membra[ne] to K^+ and Na^+ ions, D_K/D_{Na}.

If the electrode that triggers the nerve action potential extends down he whole length of a nerve axon, the whole length can be simultaneously epolarized. If, however, the initial trigger pulse is restricted to a horter length of the nerve axon, only the area adjacent to the electrode ill fire. This area of depolarization can then move down the nerve as a esult of electronic propagation. This term means simply that electric urrent flows down the longitudinal gradient of potential along the nerve nd causes a critical depolarization $\Delta\Phi$ to occur adjacent to the original epolarized area of membrane, and this critical $\Delta\Phi$ in turn triggers a iring of that region of membrane. The result is that the depolarized area preads lengthwise along the nerve, giving the propagated action potential.

The velocity of the nerve impulse was a major problem in physiology or over 200 years. Early workers referred to it as the "speed of thought". n 1834 Johannes Müller in his standard Handbuch der Physiologie concluded hat the speed was so fast that "its actual value will probably be denied o us forever". Only 16 years later, however, his pupil Herman von Helmholtz uccessfully made the experimental measurement [10]. He simply determined he difference in latency of the response of a frog muscle when stimulation as applied at two separate points along the nerve. He measured the time two different ways, by passing a current thru a ballistic galvanometer nd by graphical recording on a moving surface. The result was about) m s^{-1}. In 1922 Gasser and Erlanger [11] introduced the cathode ray cillograph into neurophysiology and in 1928 Adrian and Bronk [12] pub-shed the first measurements on individual nerve axons.

Measurements on single axons were greatly facilitated by the suggestion J. Z. Young [13] in 1936 that the giant unmyelinated axon of the squid uld be an excellent preparation for studies of nerve conduction. These ons have diameters as large as 10^3 µm, compared to the 0.1 to 20 µm of ripheral nerves of amphibians or mammals.

It is possible to squeeze out the axoplasm (protoplasm in the interior the axon) from these giant axons and to perfuse the hollow tube with lutions containing various ions and nonelectrolytes. In this way the ectrical properties of the axolemma (axonal membrane) can be studied der different nonphysiological conditions.

Perfusion studies, mainly by Tasaki and his coworkers [14], have de-neated the ionic requirements for the excitability of membranes. The ternal medium must contain bivalent ions and the internal medium, uni-lent ions. For example, excitability of squid axons can be maintained when they are perfused with CsF solution and the external medium is a Ca solution. Addition of univalent ions such as Na^+ or hydrazinium to the ternal medium increases the action potential, but external univalent io are not essential. The conditions studied by Tasaki are indeed "nonphys logical" but the experiments indicate important basic features of the ex citability of membranes.

<u>The Hodgkin-Huxley Equation</u>

Consider in Fig. 6 a long cylindrical axon of radius a. Let the res tivity of the axoplasm be ρ and the resistivity of the external medium b much less than ρ. A thin membrane of high resistance lies at $r = a$. Th outflow of total current (axial and radial) from any segment of the axon thickness Δz must vanish,

$$\pi a^2 \Delta i_z + 2\pi a i_r \Delta z = 0$$

where i_z, i_r are the axial and radial components of total current density. From Ohm's Law $\rho i_z = \partial V/\partial z$ so that

$$\frac{a}{2} \frac{\partial i_z}{\partial z} + i_r = 0$$

$$\frac{a}{2\rho} \frac{\partial^2 V}{\partial z^2} = i_r$$

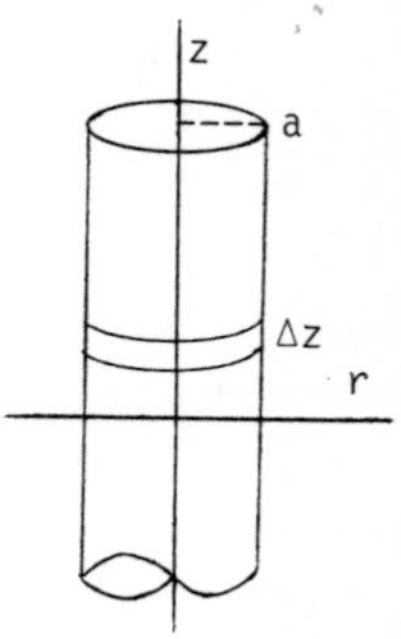

Fig. 6. Section of a nerve axon a basis for derivation of Eq. [1

The i_r consists of current density of charging the membrane $C(\partial V/\partial t)$, whe C is the membrane capacitance per unit area, and a conduction current de sity due to ions that can permeate the membrane.

In the case of squid axon, the principal permeant ions are Na^+ and Hodgkin and Huxley [14] expressed the current density as

$$i_r = C\frac{dV}{dt} + g_{Na}(V-V_{Na}) + g_K(V-V_K) + g_L(V-V_L). \quad [1$$

Here g_L represented an overall "leakage conductance" due to other ionic species. The V_{Na}, V_K and V_L are equilibrium potentials.

We know that for a given nerve under fixed conditions, the form of th action potential and the conduction velocity v are always fixed, hence

$$V(z,t) = V(z \pm vt)$$

and the wave equation applies to V,

$$= \frac{1}{v^2} \frac{\partial^2 V}{\partial t^2} \quad [17]$$

m [16] we obtain the famous Hodgkin-Huxley equation:

$$\frac{}{2} \frac{\partial^2 V}{\partial t^2} = C \frac{\partial V}{\partial t} + g_{Na}(V-V_{Na}) + g_K(V-V_K) + g_L(V-V_L) \quad [18]$$

When this equation is solved numerically only one value of the con-tion velocity v yields a finite waveform. This calculated v is further-e in excellent agreement with the experimental value. Thus the Hodgkin-ley formulation, given appropriate functions for the conductances g er voltage clamp, will also yield the action potential, its shape and duction velocity.

Excitable Membrane

Ever since the fundamental paper of Hodgkin and Huxley (1952) students research workers have been forced to think of excitation of the nerve brane in terms of the concepts they introduced. The essential concep-l novelty of the H-H formulation was the introduction of conductances ,t) that are functions of time and of potential difference across the brane. The excitation of the membrane arises from the strong nonlinear endence of the g's on the potential difference V. One might ask, refore, why the fluxes were written in a format $g(V-V_i)$ with an explicit ear dependence on V. The reason is that the g's are not instantaneous ctions of V. If V is changed abruptly, the g does not respond imme-tely, but a rate process occurs so that g changes with time to a new ue.

Hodgkin and Huxley represented the g's as powers and products of ntities that satisfy first-order differential equations (with time t as independent variable). Thus,

$$= \bar{g}_{Na} m^3 h \qquad g_K = \bar{g}_K n^4 k \quad [19]$$

e $\bar{g}_{Na}$ and $\bar{g}_K$ are the maximum values of the conductances. Each of the ameters, m, h, n, k is represented by an equation of the same form:

$$dt = \alpha (1-y) - \beta y \quad [20]$$

where y represents m, h, n, or k. The coefficients α and β are functions of the transmembrane potential difference V:

$$\alpha_m(V), \quad \beta_m(V), \quad \alpha_h(V), \quad \beta_h(V),$$

$$\alpha_n(V), \quad \beta_n(V), \quad \alpha_k(V), \quad \beta_k(V).$$

The m parameter governs the time course of the voltage-dependent opening of the Na^+ channels and the h parameter controls the spontaneous closing of these channels. The n and k parameters play an analogous role for the K^+ channels.

Under conditions of voltage clamp, V = constant, and hence $\alpha(V)$ and $\beta(V)$ are constants. Hence, integration of Eq. [20] yields

$$y = y_o e^{-(\alpha+\beta)t} + \frac{\alpha}{\alpha+\beta}(1-e^{-(\alpha+\beta)t}) \qquad [21$$

where y_o is the value of y at t = o. As t approaches infinity y approach the steady state value $\underline{y}$, so that $\underline{y} = \alpha/(\alpha+\beta)$. We can write $(\alpha+\beta)^{-1} = \tau$, time constant. Thus Eq. [20] can be rewritten as

$$\frac{dy}{dt} = \frac{\underline{y}-y}{\tau} \qquad [22$$

There will be one equation of this form for each of the parameters, m, h, and k.

The time constant for k, the spontaneous closing of the potassium channels, was not included in the original Hodgkin-Huxley formulation but was added later by Gilbert and Ehrenstein [15]. The τ_k is much longer th the τ_h for closing the Na^+ channels, being about 10 s as compared to 10 m at normal resting potentials.

The parameters of the Hodgkin-Huxley equations for normal squid axon were summarized by Cole in the convenient graphical form shown in Fig. 7. Explicit algebraic expressions for the various α's and β's as functions o V are given in the original Hodgkin-Huxley paper for squid axon. The ti constants and other parameters of nerve conduction will depend upon the temperature and upon the particular variety of nerve being studied. For example, parameters for squid axon would not be applicable to nodes of Ranvier of a frog nerve.

The student who wishes to study the literature of nerve electrochem istry must become familiar with these unusual - not to say bizarre - functi

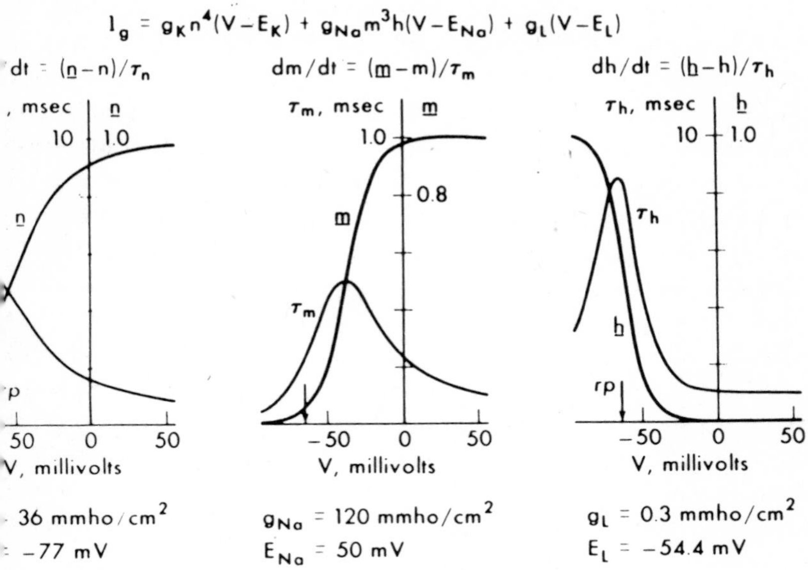

7. Variation with transmembrane potential difference V of the Hodgkin-ey parameters for normal squid axon (K.S. Cole, Membranes, Ions and lses, Univ. California Press, Berkeley, 1968).

often elegant and precise results of electrophysiologists are usually reported as measured experimental data, but rather in terms of quan-es derived from special theoretical constructs such as Goldman ratios embrane permeabilities and Hodgkin-Huxley conductances g. The dif-lties may be understood when we realize that the H-H formulation of e conduction involves at least 6 empirical functions such as appear in [20] and 21 empirical parameters.

ctrical Double Layers

The conceptual problem became even more serious when it was discovered there is in reality no general theoretical relation between nerve mem-ne potential and membrane conductance, because charges at the membrane sur- may have a large influence on the actual electric field in the membrane.

The investigation of surface charges on nerve membranes was inspired by a series of papers based on the discovery by Narahashi [16] in 1963 that squid axons could remain excitable even when the transmembrane potential V was reduced to zero. The axons were perfused with solutions in which the concentrations of K^+ ions was lowered while osmolality was maintained by addition of sucrose to the perfusing solution. Under these conditions the ionic strength of the internal medium was decreased, and in accord with Goüy-Chapman theory, the width of the electric double layer at the membrane surface was increased. The resulting situation is shown in Fig. 8. Under these perfusion conditions V across the nerve membrane migh be reduced even to zero, while $\Delta\Phi$, which determines the actual electric field in the membrane, would still maintain a relatively high value.

These experiments showed clearly that $\Delta\Phi$ and not V is the physically important potential difference in nerve conduction. The Hodgkin-Huxley formulation thus became less satisfactory as a basis for detailed molecul models of the excitation process. The properties of a membrane might indeed be expected to depend on the electric field E(x) in the membrane, bu the relation of the field to the transmembrane potential is complicated b the effects of charges on the membrane surfaces and in the electrical double layers extending from the surfaces into the surrounding media. We may note that V between the two bulk solutions is to a good approximation measurable potential difference - i.e., the difference in electrochemical potential between two phases of the "same" composition [in this case roug similar ionic composition]. The potential difference $\Delta\Phi$, however, is a typical Galvani p.d. which is not accessible to direct physical measureme as has been pointed out by Gibbs, Guggenheim, Lange and many others [17].

A quantitative estimation of the density of surface charges based up the Goüy-Chapman theory was provided by Chandler, Hodgkin and Meves [18]. For simplicity they assumed that the surface charges were only on the inside of the membrane. For a 1-1 electrolyte solution of bulk concentrat c, the potential Φ at a distance x from the interface is given by the Poisson-Boltzmann equation

$$\frac{d^2\Phi}{dx^2} = \frac{2ec}{\varepsilon_o\varepsilon} \sinh \frac{e\Phi}{kT} \qquad [2$$

here ε is the dielectric constant of the (assumed uniform) medium and ε_o the permittivity of space.

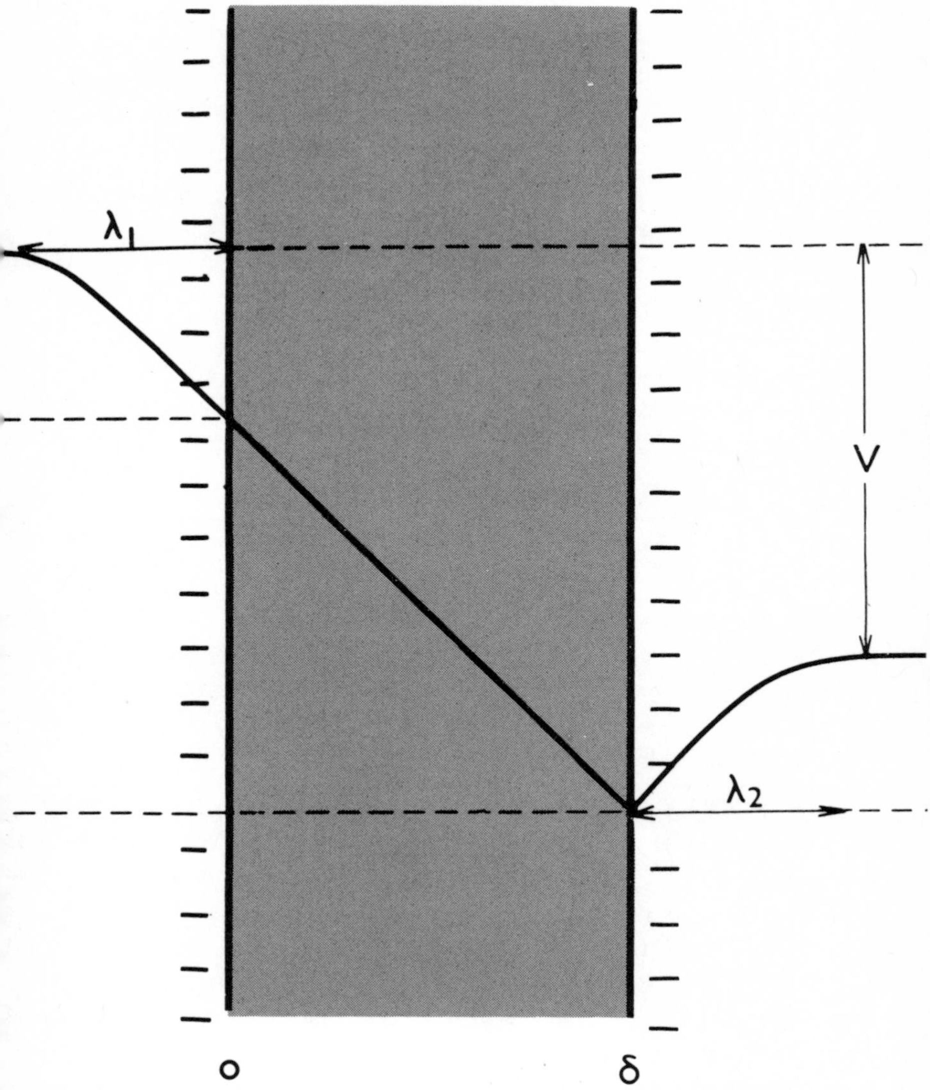

g. 8. Representation of the fixed negative charges on the surfaces of a rve membrane, and the electric potentials in the diffuse double layers. te that the potential difference ΔΦ across the membrane may differ greatly om the p.d. V between points in cytoplasm and extracellular fluid.

Integration of Eq. [23] with the boundary condition $d\Phi/dx = 0$ at $x = \infty$ ves

$$\frac{d\Phi}{dx} = - \frac{2kT}{e\lambda} \sinh \frac{e\Phi}{2kT}$$ [2

where $\lambda = \left(\frac{\varepsilon_o \varepsilon kT}{2ce^2}\right)^{1/2}$ is the Debye length or effective thickness of the double layer.

We can now obtain the charge per unit area σ at the membrane surface. At the interface between the axoplasm and the membrane (m)

$$\varepsilon_o \varepsilon' E_m - \varepsilon_o \varepsilon E_\delta = \sigma$$ [2

ε' is the dielectric constant of the membrane. If the field in the membrane is constant, $E_m = -\Delta\Phi/\delta$, and $E_\delta = -(d\Phi/dx)_{x=d}$, so that Eq. [25] becomes

$$\varepsilon_o \varepsilon'(\Delta\Phi/\delta) + \frac{2\varepsilon_o \varepsilon kT}{e\lambda} \sinh \frac{e\Phi(d)}{2kT} = \sigma$$ [2

A value of $\sigma = 1.4 \times 10^{13}$ electronic charges/cm^2 was required to fit the data on the field dependence of the sodium conductance in the perfuse squid axons. This σ corresponds to a spacing of only 27 nm between charg More detailed theories have considered the effect of restricting the negative surface charges to areas localized around the gates of the ionic channels in the membrane.

Titratable Surface Groups and Membrane Potential

The changes at the surface of the membrane arise from ionizable grou such as RCOOH, RNH_3^+ or RPO_3H_2, each with a characteristic pK value. The general problem is an interesting and difficult one. The ionization of t acid groups on the surface of the membrane depends upon the local concentration of hydrogen ions, and this depends upon the electric potential in accord with a Poisson-Boltzmann type of equation. The electric potential itself depends upon the surface charge. The potential differences $\Delta\Phi$ across the membrane should therefore depend upon the pH. Thus the surfac charges on the inner and outer interfaces of the membrane are coupled wit the transmembrane potential.

A partial solution to this problem was given by Nelson, Colonomos an McQuarrie [19]. They consider three Poisson-Boltzmann equations, one for each region in Fig. 8. Thus they employ an equilibrium theory, and consid that the equilibrium structure of the electric double layer is not affect by the ionic fluxes.

$$\frac{\Phi_1}{x^2} = \frac{-\rho_1(x)}{\varepsilon_o \varepsilon} \qquad x < 0$$

$$\frac{\Phi_2}{2} = 0 \qquad 0 \leqslant x \leqslant \delta \qquad [27]$$

$$\frac{\Phi_3}{2} = \frac{-\rho_3(x)}{\varepsilon_o \varepsilon} \qquad x > \delta$$

te that the space charge within the membrane is assumed to be zero.

A solution to this problem for the case corresponding to the Goldman nstant-field model within the membrane was obtained. The coupling term ween the two surface potentials depends upon ε'/ε, the ratio of the di- ctric constants of the membrane and of the surrounding aqueous medium. n $\varepsilon'/\varepsilon << 1$ the theory reduces to that previously given by Gilbert and enstein [20] for the titration of the surface charge. For a special e in which only univalent ions and singly charged surface negative groups concerned,

$$_o/2kT\ (e^{e\psi_o/kT} - 1) = \frac{e^2 K_1\ e^{2e\psi_o/kT}}{\varepsilon_o \varepsilon S_1 kT\chi([H^+] + K_1 e^{e\psi_o/kT})} \qquad [28]$$

In this expression the transmembrane potential V is assumed to be zero. it is not zero, we should write $(\psi_o - V)$ instead of ψ_o in Eq. [28]. $[H^+]$ the concentration in the solution [cytoplasm]. K_1 is the dissociation stant of the surface acid, $AH = A^- + H^+$:

$$= [H^+]_o \frac{\alpha}{1 - \alpha}$$

hat the surface charge can be calculated from $\sigma = -e\alpha/S$, while $[H^+]_o =$ $e^{-e(\psi_o - V)/kT}$. Thus if we know S and K_1 we can calculate the potential he surface as a function of pH in the surrounding solution.

Instantaneous Currents

The Goldman-Hodgkin-Katz equation [15] was derived for the steady e currents thru an unexcited membrane. Yet we noted that Hodgkin and ey used an ohmic relation for the ionic currents thru the activated d axonal membrane.

The experimental situation in regard to the different I-E characteristics of various nerve membranes is interesting. In the case of conducti at the nodes of Ranvier in myelinated nerves of the toad Xenopus [21] both K^+ and Na^+ currents follow the GHK Eq. [15]. In the case of nodes of nerve of the frog Rana [22] the Na^+ current follows GHK but the K^+ current is ohmic. In the much studied case of the giant squid axon (an unmyelinated nerve) neither K^+ nor Na^+ current follows GHK. This behavior has led to much theoretical speculation, being called "the failure of instantaneous rectification".

The GHK equation would be linearized if in Eq. [15] $\eta << 1$. For a typical physiological situation of a nerve membrane, however, $\Delta\Phi$ is about 70 mV whereas at 300 K, kT/e is only 25 mV. Thus an ohmic conductance can not be explained as a linear limit of the GHK equation.

Consider the "instantaneous current" thru the nerve membrane, measure in the time interval immediately after application of a voltage step. The is always a capacitative term representing the charging or discharging of the membrane. After subtracting this term, we can write for each ionic species:

$$I_i = I_o f(t,E).$$

Note that the time dependent term has been explicitly separated from the total current. When $t = 0$, $f(t,E) = 1$, so that $I_i = I_o$ is called the "instantaneous current".

There are two views about the interpretation of the instantaneous currents I_o. According to the view of Noble, et al. [23], I_o represents the current thru the ionic channels in their fully opened condition. Thu it would be the current observed if some method could be devised to keep the channels open, or to block the closing mechanism. If this view is correct, then some special structural features must be postulated to explain why I_o for nodes of Ranvier follows electrodiffusion theory whereas I_o for squid axon displays an ohmic behavior. One example of such a stru tural theory was given by Woodbury [6] based upon a chain of potential-energy barriers of varying heights in the membrane channels.

A quite different view of the I_o was espoused by Bass [24]. He proposed that the difference between the functional forms of I_o for nodes a for axons reflected a difference in the response time of the membrane co ductances to a voltage step. In the case of nodes, there is a high dens channels and the ionic currents can quickly achieve the form given by ady-state electrodiffusion, before the potential-dependent gating pro- ses have time to be appreciably activated. In the case of axons, the)onse of the membrane is slower and the process of excitation supervenes)re the electrodiffusion regime is reached. No special structural :ures of the membrane need be adduced to explain the different behaviors axons and nodes.

It is rather strange that this basic question in regard to nerve con- :ion has not been answered yet. Experiments can, however, be made in :h the proteins that close the gates of the Na^+ channels are selectively ved by perfusing the axon with a proteolytic enzyme. It should thus be ible to measure ionic currents thru the nerve membrane while the mech- m that deactivates the Na^+ channels is eliminated.

ng Currents

When an axon under voltage clamp is suddenly depolarized by a step in transmembrane potential, there is an immediate capacitative current. is observed as a rapid spike which is finished almost as soon as the voltage is completed. Superimposed on this capacitative current there small current called the gating current, which represents the move- s of charges associated with opening and closing ionic channels in the rane. There is good chemical evidence, based on selective effects of king agents, that separate channels exist for the Na^+ and K^+ ions. The types of channels are not completely selective. For example, in squid it has been estimated that the K^+ permeability thru the Na^+ channel is that of Na^+.

For a long time these gating currents, though believed to exist on retical grounds, could not be detected owing to technical difficulties. beautifully designed experiment, Armstrong and Bezanilla [25] detected gating currents for the Na^+ channels. They have also been studied by es and Rojas [26].

Fig. 9 shows the gating current and the Na^+ current for a squid axon rnally perfused with 55 mM CsF + sucrose. In the external "sea-water" normal [Na^+] was replaced by [$Tris^+$].

The gating current is a transient outward current as the Na^+ channels . When the Na^+ channels close on repolarization of the membrane, a sient inward current can be detected. This inward charge transfer just ces the initial outward transfer. The evidence indicates that these ents are not due to ionic transfers thru the membranes but rather to

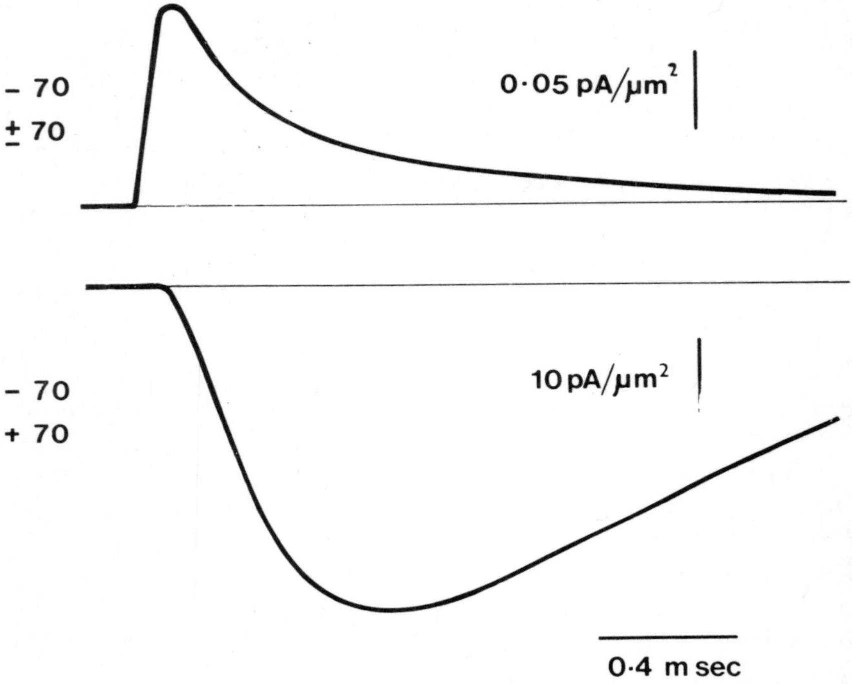

Fig. 9. The upper curve is the gating current in a squid axon recorded sea water in which $Tris^+$ has replaced Na^+, with 0.55 M CsF internally (3.5°C). The lower curve is the Na^+ current as recorded in normal sea water. The holding potential is -70 mV. Capacitative currents are eli inated by subtracting the currents measured with equal + and - pulses.

charge transfers within the membrane. Such transfers may be either cha in the orientation of dipoles or motions of internally bound ionic char particularly protons.

Random Processes and Membrane Noise Spectra

A field that has shown recent interesting developments is the stud random processes (noise spectra) in electrochemical systems. Applicati can be made to the membranes of axons or synapses. Such noise spectra sometimes be interpreted in terms of detailed mechanisms of membrane co ductance, such as opening and closing of ionic channels. Essentially t ise spectra provide a method of studying the same sorts of rate processes are measured in the relaxation processes following a sharp perturbation the resting state of a system.

Three applications of noise spectra have been of special interest in e electrochemistry of nerves.

) Noise arising from the resting membranes of nerve cells, as in squid axons or nodes of Ranvier.

) Threshold fluctuations of two types: (a) The statistics of the critical depolarization necessary to cause firing of an action potential. These data can be expressed in the form of the probability of firing as a function of the stimulating current. (b) Following delivery of a depolarizing pulse to the axonal membrane there will be a time delay or latency before the action potential ensues. The latency distribution can be expressed in the form of delay time as a function of intensity of stimulus.

Noise arising as voltage fluctuations across the postsynaptic membrane of neuromuscular junctions exposed to microiontophoretic injection of acetylcholine. Analysis of this noise has provided an indication of the interaction of individual acetylcholine molecules with receptor sites.

This entire field has been well covered by recent reviews and papers ,28] and only brief comments will be made. In the case of axonal memnes we would be most interested in any noise spectrum that may arise n random opening and closing of ionic channels. This noise, however, t be separated from that of other sources, which would exist for contance thru channels that are open. There are also instrumental sources noise that must be minimized, in particular, noise due to random proses at electrode surfaces.

The analysis of the noise spectrum is based upon the Wiener-Khintchin orem, which relates the power spectrum $S(\omega)$ of the noise to a correlation tion $C(\tau)$,

$$= \int_{-\infty}^{+\infty} S(\omega) \cos\omega\tau \, d\omega$$

A simple correlation function that is often suitable would be

$$= e^{-\tau/\tau_c}$$

where τ_c is a measure of the average time that the system exists without changing a characteristic function of its state x(t) by more than 1/e of its initial value.

The first kind of nerve membrane noise to be extensively studied was found over a range of frequencies from 10^{-1} to 10^4 Hz in the recordings of membrane voltages at nodes of Ranvier in frog nerves. Its power spectrum varied inversely as the frequency and hence it was called 1/f noise. Present evidence assigns its mechanism to fluctuations in the movements of K^+ ions thru open channels in the membranes. Another type of voltage fluctuation in frog nodes is called "burst noise". It occurs in the form of miniature depolarization pulses, and is believed to be associated with momentary dielectric breakdowns of small areas of the membrane.

More recent developments have detected noise components that may be due to random opening and closing of K^+ channels and Na^+ channels. Such noise spectra permit an estimation of the conductances of individual channels. In the case of frog nodes, for example, the value of K^+ and Na channel conductances were estimated as 4 x 10^{-12} S and 2.4 x 10^{-12} S, respectively. These estimates appear to contradict estimates based on su face density of channels, which led to a ratio of at least 10/1 for the N and K^+ conductances per channel. Somewhat similar measurements on squid axon gave 4 x 10^{-12} S for the K^+. These results suggest that the channel may be similar in nodes and axons, while the surface density of channels would be about 10 times higher in the nodes.

Molecular Mechanisms of Excitation

We have outlined some of the interesting electrochemical problems of nerve membranes, but we have said little about the detailed molecular mec anisms of the excitation process. Some theories would explain excitatio as a cooperative phenomenon (like a phase change) in the membrane, occurring at certain critical concentrations of bivalent and univalent ions (Tasaki [29]). We shall not consider these theories here. They are imp tant for certain types of synthetic ion-exchange membranes, but I believe that they will not explain nerve excitation. This belief is based upon two reasons: (1) Extensive theoretical studies by Hill and Chen [30] indicated that, at least in the case of K^+ channels, cooperative kinetics cannot be made to fit the experimental data. (2) The high electric fiel across nerve membranes is a fact of nature. The electric fish provide a direct demonstration. It seems likely that the secret of nerve excitati must be related to physical effects of such high fields and abrupt chang such fields.

There are two ways in which high fields, or rapid depolarization from n fields, may lead to membrane excitations: (1) reorientation of di-es (2) effects on ionization of weak electrolytes (Wien dissociation ects) and other shifts of bound charges.

The dipole reorientation mechanism has been popular with physiologists it appears to have serious quantitative drawbacks. The average value a dipole moment μ in the direction of a field F is

$$= \mu L(\mu F/kT)$$

re $L(x)$ is the Langevin function,

$$x) = \coth x - \frac{1}{x}$$

Consider a field of 10^5 V/cm which is reduced suddenly by 1/3 of that ue. If μ = 10 Debye, $\bar{m}/\mu$ is reduced from 0.1395 to 0.0732. This change d correspond to an average rotation of about 4° away from the field ection. Such an effect seems too small to provide a reliable gating anism unless it is amplified by some change in conformation of membrane ecules. If μ is made much larger, the resultant dipoles would extend ss much of the membrane thickness and changes in orientation of such les would be quite slow. The time constant of the observed gating cur-s is about 0.1 ms, so that even quite slow reorientations would be ssable.

Instead of moving large molecules about in the membrane it seems more onable, however, to suppose that changes in the electric field act by ing shifts of small bound ions, especially protons. A mechanism in-ing such proton shifts was suggested by Kirkwood and Shumaker [31] for rization of proteins in an electric field and a consequent contribution he dielectric loss of protein solutions. Actually such a loss mech-m has not been unequivocally detected in protein solutions. Hill and [32] have discussed its relevance to nerve excitation.

The Kirkwood-Shumaker mechanism is based on the equilibrium distri-on of counterions, particularly protons, in an electric field. It ly imposes an electrostatic energy term upon the distribution of ener- of binding sites in the absence of the field. It does not include an al field effect on the dissociation of energy of the type described by as the "field dissociation effect" in weak electrolytes.

A theory of the Wien effect was given by Onsager [33]. The Wien effect would be particularly suited to play a role in nerve excitation since it is a strongly nonlinear function of the field strength. Its po sible role in nerve conduction appears to have first been mentioned by C [34] in 1965, but he rejected it on the grounds that it would require a pendence of ionic conductance on the absolute value of the field strengt $|E|$. In 1969 Gilbert and Ehrenstein [35] showed that $g_K(V)$ does not increase continuously with depolarization but eventually falls again to yi a g vs V curve that is roughly symmetrical about a value of V that depen on the external Ca^{2+} concentration. Thus a possible role for the Wien effect became more likely.

Bass and Moore [36] have given a theoretical model for nerve excitation based on the Wien effect. McIlroy [37] has shown how the voltage-clamp data of Hodgkin and Huxley on squid axon can be interpreted by a W effect model. The particular model used by McIlroy was based on field c pendent enzymatic reactions in the membrane (as in theories of Nachmansc [38]) but a similar fitting of the data could be obtained in terms of cc formation changes of gate proteins.

My present belief is that field-dependent proton shifts in nerve me branes provide the most reasonable basis for triggering the action poten The proof of this model will require identification of the molecules inv ved in the gating process, independent measurement of the field depende of their conformation and the mechanism by which it is altered thru prot shifts, and finally reconstitution of an excitable membrane with the exc tation properties of naturally occurring axons and nodes of Ranvier. Me while experimental work on nerve membranes would profit by a greater emp sis on the measurement of surface charge distributions and effects of pH and by careful data on deuterium isotope effects designed to elucidate t roles of proton transfers in the excitation process.

The electrochemist who wishes to enter this field would usually nee to begin work in collaboration with an experienced neurophysiologist. I would be difficult to learn how to handle the experiments on nodes and axons simply by reading the published literature.

The basic problems of the molecular mechanisms of nerve conduction synaptic transmission may well be solved within the next 10 years.

The structure of nerve networks and the internal logic of central nervous systems is now a more difficult problem in an early stage of exp ation. Electrochemistry will have an important part in elucidating the mechanism by which nerve networks store information. The question here w the pulses of electrical depolarization that are produced by trans-ction of sensory input data are transcribed into a permanent record in e brain. Physical chemists may tend to believe that this question will answered by an electrochemical mechanism that in some way couples mem-ane depolarization or ionic fluxes to alterations in rates of synthesis particular biological macromolecules. Some natural philosophers, wever, would not accept this faith in reductionism. They would contend at memory is a function not of the brain but of the mind, an entity that y be beyond even the reach of physical chemistry [39]. We would still, wever, have the problem of how the mind influences the brain. Bass [40] s published a provocative theory about this, which, altho it is unlikely have solved the problem, may excite both the minds and neurons of its aders.

FERENCES

. T. Szabo and A. Fessard, Handb. Sensory Physiol., III/3(1974)59.
. C.M. Armstrong, Quart. Rev. Biophys., 7(1975)179.
. M. Planck, Ann. Physik Chem., 39(1890)161.
. D.E. Goldman, J. Gen. Physiol., 27(1943)37.
. R.B. Parlin and M. Eyring, in H.T. Clarke (Editor), Ion Transport Across Membranes, Academic Press, New York, 1954.
. J.W. Woodbury, Adv. Chem. Phys., 21(1971)601.
. R. Schlögl, Z. Physik. Chem., N 1(1954)305.
. A.L. Hodgkin and B. Katz, J. Physiol. (London), 108(1949)37.
. K.S. Cole, Arch. Sci. Physiol., 3(1949)253.
. H.v. Helmholtz, Arch. Anat. Physiol., 71-73(1850)276.
. H.S. Gasser and J. Erlanger, Am. J. Physiol., 62(1922)496.
. E.D. Adrian and D.W. Bronk, J. Physiol. (London), 65(1928)81.
. J.Z. Young, Quart. J. Micr. Sci., 78(1936)367.
. A.L. Hodgkin and A.F. Huxley, J. Physiol. (London), 117(1952)500.
. D.L. Gilbert and G. Ehrenstein, Biophys. J., 9(1969)447.
. T. Narahashi, J. Physiol. (London), 169(1963)91.
. E.A. Guggenheim, in F.G. Donnan and A. Haas (Editors), A Commentary on the Scientific Writings of J. Willard Gibbs, Vol. 1, Yale University Press, New Haven, 1936, pp. 181-211, 331-349.
3. W.K. Chandler, A.L. Hodgkin and H. Meves, J. Physiol. (London), 180(1965)821.

19. A.D. Nelson, P. Colonomos and D.A. McQuarrie, J. Theor. Biol., 50(1975)317.
20. D.L. Gilbert and G. Ehrenstein, J. Gen. Physiol., 55(1970)822.
21. F.A. Dodge and B. Frankenhauser, J. Physiol. (London), 148(1959)188.
22. F.A. Dodge and B. Frankenhauser, J. Physiol. (London), 143(1958)76.
23. D. Noble and R.W. Tsien, J. Physiol. (London), 200(1969)205.
24. L. Bass, J. Theoret. Biol., 48(1974)133.
25. C.M. Armstrong and F. Benzanilla, Nature (London), 242(1973)459.
26. R.D. Keynes and E. Rojas, J. Physiol., 239(1974)393.
27. A.A. Verveen and L.J. DeFelice, Prog. Biophys. Mol. Biol., 28(1974)1
28. F. Conte and E. Wanke, Quart. Rev. Biophys., 8(1975)451.
29. I. Tasaki, in C. Thomas (Editor) Nerve Excitation, Springfield, IL, 1968.
30 T.L. Hill and Y. Chen, Proc. Nat. Acad. Sci. U.S.A., 68(1971)1711, 24
31. J.G. Kirkwood and J.B. Shumaker, Proc. Nat. Acad. Sci. U.S.A., 38(1952)855, 863.
32. T.L. Hill and Y. Chen, Proc. Nat. Acad. Sci. U.S.A., 66(1970)607.
33. L. Onsager, J. Chem. Phys., 2(1934)599.
34. K.S. Cole, in T.H. Waterman and H.J. Morowitz (Editors), Theoretical and Mathematical Biology, Blaisdale Publishing Co., New York, 1965, pp. 136-171.
35. D.L. Gilbert and G. Ehrenstein, Biophys., 9(1969)447.
36. L. Bass and W.J. Moore, in A. Rich and N. Davidson (Editors), Structural Chemistry and Molecular Biology, Freeman, San Francisco, 1968, p. 356.
37. D.K. McIlroy, Math. Biosci., 7(1970)313.
38. D. Nachmansohn, Biochimie, 55(1973)365.
39. W. Penfield, The Mystery of the Mind, Princeton University Press, 197
40. L. Bass, Foundations of Physics, 5(1975)159.

HEORY AND APPLICATIONS OF ELECTRON TRANSFERS AT ELECTRODES AND IN SOLU-ION

. A. MARCUS

epartment of Chemistry, University of Illinois, Urbana, Illinois 61801

3STRACT

The theory of simple electron transfers at electrodes and in solution reviewed, and various thermal fluctuations of coordinates leading to ectron transfer are described. A simplified derivation is given of the ee energy of such fluctuations. Implications of the theoretical equa-ons for experiment are described, some of the relevant results having en summarized recently in Dahlem Konferenz Phys. Chem. Sci. Res. Rept., 1975)477. They include relations between rate constants of cross-actions and self-exchange reactions, between rates of reactions at elec-odes and those in solution, nonspecific solvent effects, chemilumines-nce, and other properties. Approximate equations of the BEBO type are so given for Tafel slopes, Bronsted slopes and rates of cross-reactions, r systems involving rupture and formation of bonds.

TRODUCTION

The theory of reactions at electrodes has much in common with that reactions in solution. The electrode behaves as one large reactant, with special properties: it has numerous electronic energy levels and energy of those levels is controllable by the electrode potential. t as one reactant in solution may bind the other, the electrode may orb the other reactant.

In the present paper we shall be concerned principally with simple ctron transfer reactions at electrodes and in solution, reactions ch involve no rupture or formation of chemical bonds. Elsewhere [1] have discussed reactions involving rupture of a chemical bond at an ctrode M, e.g., Eqs. [1]-[2], and for brevity omit a discussion of this ic here.

$$H_3O^+ + M(e) \rightarrow H_2O + H - M \qquad [1]$$

$$H_3O^+ + H - M \rightarrow H_2O + H_2 + M^+ \qquad [2]$$

A simple electron transfer reaction at an electrode can be written as

$$A_1(ox) + M(n\bar{e}) \rightarrow A_1(red) + M \quad , \qquad [3]$$

while a simple electron transfer between species A_1 and A_2 in solution i[s] represented as

$$A_1(ox) + A_2(red) \rightarrow A_1(red) + A_2(ox) \quad , \qquad [4]$$

where (ox) and (red) denote the oxidized and reduced forms of the two chemical species A_1 and A_2. The ionic charges on these reactants are usually about one to three, though sometimes as high as four and sometimes as low as zero. The solvation energies are therefore very large and their fate during the course of the reaction must be analyzed carefully.

POTENTIAL ENERGY CURVES, SURFACES, FLUCTUATIONS, AND RATES

It is useful to examine first the vibrational motion within the reactant in Eq. [3], considering initially the case of one vibrational coordinate q. (The same plot suffices for Eq. [4] also.) A plot of the potential energy U versus q for the system on the left hand side of Eq. [3] is labelled R in Fig. 1, and a plot of U versus q for the system on the right hand side of Eq. [3] is labelled P [2,3].

There are many electronic states of the metal--a continuum of them. Thus, Fig. 1 for the reaction in Eq. [3] should consist of many paralle[l] curves, vertically displaced from each other, one for each electronic state. However, one can show [3,4] that the electrons donated from the electrode M to A(ox) in Eq. [3] come from levels near the Fermi level, and most of them donated from A(red) to M in the reverse reaction go in levels near the Fermi level. Thus, it suffices to confine our attentio[n] for the present purposes to the two curves in Fig. 1 and take the energ[y] level of the electron in the metal as the Fermi level. The following analysis then applies equally well to reactions [3] and [4].

Several facts are noted:

(1) The minima of the two curves occur at different values of q, [re]flecting the fact that the equilibrium bond length in A_{ox} is different (usually shorter in the case of transition metal io[ns] and a metal-ligand bond) from that in A_{red}.

(2) The relative height of the two minima ΔU^0 depends on the electrostatic potential, the P curve being lowered vertically

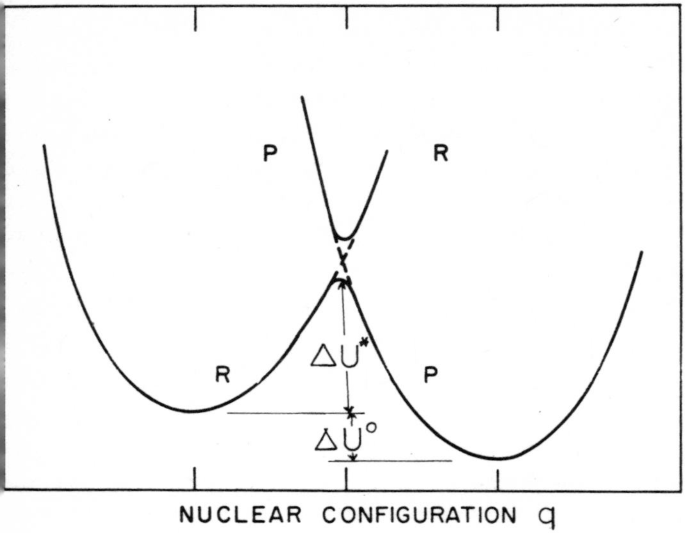

g. 1. A plot of the potential energy U of the system consisting of actants plus solvent (R), along some coordinate q, and of the system nsisting of products plus solvent (P), holding all other coordinates xed, for reaction [3] or [4].

relative to the R curve by making the electrode more negative, i.e., by decreasing $e(\varphi_M-\varphi_S)$ where the φ's represent the potentials of the metal and solution.

(3) When the difference of equilibrium bond lengths Δq^0 is less, for a given ΔU^0, the barrier height ΔU^* is less.

(4) When the P curve is lowered relative to the R one, the barrier height ΔU^*, corresponding to the intersection of the R and P curves, is reduced.

The two curves have further properties: When the reactant is far om the electrode the appropriate curves in Fig. 1 merely cross in the ntersection region (dotted lines there). When the reactant is close to e electrode, the electronic interaction of the orbitals of the reactant nd the electrode perturbs the dotted line curves, particularly near the int of intersection. At that point the unperturbed R and P electronic antum states are degenerate, and the degeneracy is broken by the interction. The new curves are the solid curves. The energy of the maximum f the lower solid curve is less than that at the intersection by an ectronic interaction energy ε_{12}, the "resonance energy."

When ε_{12} is zero (e.g., large separation distance between electrode and reactant) a system having a potential energy curve R will undergo thermal fluctuations of the vibrational coordinate q, but not transfer to the P curve even on passing through the intersection region. When ε_{12} is nonzero, there is a coupling of the electronic orbitals of the reactant and the electrode, and there is a nonzero probability κ of going fro the R curve to the P curve when the system passes through the intersection region. When ε_{12} is small the reaction has a small κ and the reaction is termed nonadiabatic. When κ is of the order of unity the reacti is adiabatic. The electronic factor κ can be calculated by the Landau-Zener-Stueckelberg formula [5], or by a more numerical quantum mechanica computation [6], if ε_{12} and the properties of the R and P curves are kno

One notes from Fig. 1 that a fluctuation of coordinate q is needed for the electron transfer to occur: Not only must the reactant-electrod distance, or in Eq. [4] the reactant-reactant distance, become sufficient small for electronic orbital overlap to occur but also the transfer does not occur at the equilibrium value of q for the R-curve. Only by absorp tion of light could the system move from the R curve to the P curve at this q. Otherwise the vibrational momentum p, given by $\sqrt{2m(E-U)}$, would not be conserved during the electronic transition and thus the Franck-Condon principle would be violated. However, at the crossing point this momentum p is substantially conserved, so the transfer from the R curve t the P curve can occur.

A simple calculation provides an estimate of the barrier height ΔU^* in Fig. 1: It will be assumed that the resonance energy ε_{12} is small enough that the height of the crossing-point is approximately that of th maximum of the lowest solid curve in Fig. 1. We let the potential energ of the R and P curves be approximated by

$$U^r(q) = \frac{1}{2} k(q-q_0{}^r)^2 \qquad [5]$$

$$U^p(q) = \frac{1}{2} k(q-q_0{}^p)^2 + \Delta U^0 \quad , \qquad [6]$$

where $q_0{}^r$ and $q_0{}^p$ are the equilibrium bond lengths for the A(ox) and A(red), respectively. ΔU^0 is negative in Fig. 1, and is a linear function of $ne(\varphi_m-\varphi_s)$ in the case of reaction [3]. At the intersection of the dotted curves we have

$$U^r = U^p \quad , \qquad [7]$$

an equation which can be solved for the value of q, $q^\ddagger$ at the intersecon. Introducing this $q^{\ddagger}$ into Eq. [5] one obtains ultimately

$$\Delta U^* = (\lambda_i/4)[1 + (\Delta U^0/\lambda_i)]^2 \quad , \qquad [8]$$

ere

$$\lambda_i = \frac{1}{2} k(\Delta q^0)^2 \quad , \qquad \Delta q^0 = q^{0p} - q^{0r} \quad . \qquad [9]$$

/4 is actually the barrier height when the ΔU^0 in Fig. 1 is zero.

The rate constant k_r would be given by the collision frequency Z per it area of electrode per unit time in the electrode case and per unit ncentration per unit time in the homogeneous case, multiplied by the ltzmann factor $\exp(-\Delta U^*/RT)$, by a Boltzmann factor $\exp(-w^r/RT)$ for inging the reactant to the electrode, and by $\varkappa$:

$$k_r = \varkappa Z \exp[-(w^r + \Delta U^*)/RT] \quad . \qquad [10]$$

re, w^r is the work term, if any, required to bring the reactants to- ther to some separation distance R. When such work terms occur, the ^{0} in Eq. [8] is replaced by $\Delta U_R{}^0$, the ΔU at that separation distance R. us, Eq. [8] is replaced by

$$\Delta U^* = (\lambda_i/4)[1 + (\Delta U_R{}^0/\lambda_i)]^2 \qquad [11]$$

$$\Delta U_R{}^0 = \Delta U^0 + w^p - w^r \quad , \qquad [12]$$

ere ΔU^0 is the ΔU at infinite separation and w^p is the work to bring e products together to the separation distance R.

The above calculation of ΔU^* is classical, and indeed classical chanics is commonly used nowadays to treat reactive collisions, by cal- lating the trajectories of all the atoms of a reacting molecule or pairs molecules during a collision. However, for some problems a quantum chanical treatment is needed, for example for treating a protonic vibra- on in Eq. [1]. A quantum treatment is given in Refs. [7,8], other ap- oximations being introduced as well.

The results in Eqs. [10]-[12] were derived for the case of one vi- ation. The derivation can be extended to all the vibrations of the re- tant(s) in Eq. [3] or [4]. The potential energy curves of Fig. 1 are placed by potential energy surfaces, plotted as a function of all the s in the system rather than just one. If there are N q's, the inter- ction which is a point in Fig. 1 becomes a hypersurface of N-1 dimen- ons in the N-dimensional case. The coordinate q in Fig. 1 then repre- nts some path in the N-dimensional q-space and the R- and P-curves are ofiles of the actual potential energy surfaces plotted along that path.

In effect, λ_i becomes a sum of terms of the type in Eq. [9], summed over all vibrations, i.e.,

$$\lambda_i = \frac{1}{2}\Sigma_i k_i(\Delta q_i{}^0)^2 \quad , \qquad [1$$

and k_i is related to the force constants of a bond $k_i{}^r$ and $k_i{}^p$ in the oxidized and reduced forms [3]. A somewhat more sophisticated treatment of λ_i is given in Ref. [3], leading to k_i being a certain average of k_i and $k_i{}^p$.

There remain the solvent fluctuations outside of the inner coordination shell of the reactant in Eq. [3] or reactants in Eq. [4]. Here, t potential energy functions do not depend on the solvent coordinates (orientations, translations) in the simple quadratic fashion in Eqs. [5] and [6] and of Fig. 1. The treatment of the solvent coordinates is correspondingly more complicated. However, one feature is immediately clear: Just as a thermal fluctuation of vibrational coordinates was needed to reach the intersection region in Fig. 1, a suitable thermal fluctuation of solvent orientation coordinates or reactant's vibrations also permits the system to reach the N-1 dimensional hypersurface (the intersection region). A statistical mechanical treatment of the free energy associated with these fluctuations is given in Ref. [3]. Dielectric continuum theory also permits an estimate of the latter to be made, and was first given for the homogeneous reaction case in 1956 [9], and in ONR Technical Report No. 12 (presented elsewhere in this volume) in 1957 for the electrode reaction, Eq. [3].

The intersection region in Fig. 1, now for a generalized coordinate q and a many-dimensional configuration space, is reached by a therm fluctuation of coordinates q from their most probable values. Such a fluctuation produces a corresponding thermal fluctuation in the solvent dielectric polarization function. In the intersection region this polarization is appropriate neither to the charges of the reactants nor to tho of the products but rather to some compromise, which depends on ΔU^0 or mo generally on a free energy of reaction $\Delta G^{0\prime}$ or in the electrode case on half-cell potential E minus the standard potential $E^{0\prime}$, $E-E^{0\prime}$. This dielectric polarization was termed nonequilibrium polarization [9] and indeed the original dielectric continuum (nonstatistical mechanical) derivation [9,10] was concerned with fluctuations in solvent dielectric polarization.

It is instructive to give a somewhat simpler derivation of the free

rgy change needed to reach the intersection region by these fluctua-
ns of solvent dielectric polarization. This derivation is given in
next section.

1PLIFIED DERIVATION OF NONEQUILIBRIUM POLARIZATION EXPRESSION

We consider the homogeneous reaction system Eq. [4] first and then licate the modification for the electrode case Eq. [3]. We denote the rges by e_i and radii by a_i for reactants 1 and 2 ($i = 1,2$), and add a erscript p to the e_i to denote charges of the products. The static lectric constant of the solvent medium is denoted by D_s and the square the refractive index (the "optical dielectric constant") by D_{op}. The aration distance is denoted by R.

A nonequilibrium dielectric polarization of the medium can be pro- ed in a reversible manner by a two-step charging process. Since each p is reversible, the free energy of formation of this nonequilibrium tem, i.e., the free energy of this polarization fluctuation, can be culated in a relatively straightforward manner. The two-step charging cess is the following, at a given separation distance R.

(1) Change the charge of each reactant i from e_i to e_i', e_i' being so chosen to produce the desired orientational-vibrational dielectric polarization.

(2) Change the charge of each particle i back from e_i' to e_i, holding the above orientational-vibrational dielectric polarization fixed.

details of the calculation are as follows, where the electrostatic ential in the solvent medium at any point $\underset{\sim}{r}$ is denoted by $\psi(\underset{\sim}{r})$.

p 1

The value of e_i and $\psi(\underset{\sim}{r})$ at any stage ν of the charging process are oted by e_i^{ν} and $\psi^{\nu}(\underset{\sim}{r})$, respectively. They are given by

$$\psi^{\nu}(\underset{\sim}{r}) = \frac{e_1^{\nu}}{D_s r_1} + \frac{e_2^{\nu}}{D_s r_2} \qquad [14]$$

$$e_i^{\nu} = e_i + \nu(e_i' - e_i) \quad , \qquad [15]$$

re r_i is the distance from the field point $\underset{\sim}{r}$ to the center of ion i. aries from 0 at the beginning of the charging process to 1 at the end, and e_i^{ν} can be written, thereby, as in Eq. [15].

The potential at the surface of ion 1 due to the medium and to ion is obtained by replacing r_1 by a_1 in Eq. [14]. Φ_1^{ν}, the potential there minus the self-potential, is obtained by subtracting e_1^{ν}/a_i from [14]:

$$\Phi_1^{\nu} = \frac{e_2^{\nu}}{D_s r_2} + \frac{e_1^{\nu}}{a_1}\left(\frac{1}{D_s} - 1\right) \quad . \qquad [16]$$

The average of Φ_1^{ν} over the surface of ion 1 is denoted by $\bar{\Phi}_1^{\nu}$ and is found to be

$$\bar{\Phi}_1^{\nu} = \frac{e_2^{\nu}}{D_s R} + \frac{e_1^{\nu}}{a_1}\left(\frac{1}{D_s} - 1\right) \ . \qquad [17]$$

The average leading from Eq. [16] to [17] will be recognized as the well known electrostatic result [11] that the average value of a $1/r_2$ potential from a uniform distribution over a sphere is $1/R$.

When $\bar{\Phi}_1^{\nu}$ is multiplied by an increment of charge $d(e_1^{\nu})$, i.e., $(e_1' - e_1)d\nu$, and integrated over ν from 0 to 1 and when the same integration is performed for ion 2, and both terms summed we obtain the work term W_I required in charging step 1:

$$W_I = \int_0^1 \bar{\Phi}_1^{\nu}(e_1'-e_1)d\nu + \int_0^1 \bar{\Phi}_2^{\nu}(e_2'-e_2)d\nu \quad . \qquad [18]$$

Eqs. [17]-[18] yield immediately

$$W_I = \frac{e_1\Delta e_2 + e_2'\Delta e_1}{D_s R} + [e_1\Delta e_1 + \tfrac{1}{2}(\Delta e_1)^2]\left(\frac{1}{D_s} - 1\right)\frac{1}{a_1} + [e_2\Delta e_2 + \tfrac{1}{2}(\Delta e_2)^2]\left(\frac{1}{D_s} - 1\right)\frac{1}{a_2} \qquad [19]$$

where

$$\Delta e_i = e_i' - e_i \quad . \qquad [20]$$

When e_1 and e_2 are both zero, the $1/R$ term becomes the usual coulomb repulsion $e_2'e_1'/D_sR$, the $1/a_1$ term the well-known Born charging term for ion 1 $(e_1'^2/2a_1)[(1/D_s)-1]$, and the $1/a_2$ term the Born charging term for ion 2.

Step 2

The charges are now given by Eq. [21], where ν goes from 0 to 1.

$$e_i^{\nu} = e_i' + \nu(e_i - e_i') \quad . \qquad [21]$$

If $\psi_I(\underset{\sim}{r})$ denotes the potential at the end of step 1 and $\psi^{\nu}(\underset{\sim}{r})$ the potential at any state ν of step 2, the change of potential during step 2, for any ν, is $\psi^{\nu}(\underset{\sim}{r}) - \psi_I(\underset{\sim}{r})$. Since the medium responds to a change of charge

$_i^\nu$ only via the optical dielectric constant D_{op}, now, we have during ep 2

$$\delta\psi^\nu(\underset{\sim}{r}) = \frac{\delta e_1^\nu}{D_{op}r_1} + \frac{\delta e_2^\nu}{D_{op}r_2} \quad . \qquad [22]$$

iting δe_i^ν as $\nu(e_i - e_i')$ and $\delta\psi^\nu$ as $\psi^\nu - \psi_I$ we have

$$\psi^\nu(\underset{\sim}{r}) = \psi_I(\underset{\sim}{r}) + \frac{\nu(e_1 - e_1')}{D_{op}r_1} + \frac{\nu(e_2 - e_2')}{D_{op}r_2} \qquad [23]$$

ere

$$\psi_I(\underset{\sim}{r}) = \frac{e_1'}{D_s r_1} + \frac{e_2'}{D_s r_2} \quad . \qquad [24]$$

$^\nu$, the average potential on the surface of ion 1 minus the self-poten- al is obtained by subtracting e_1'/r_1 from Eq. [24] and $\nu(e_1-e_1')/r_1$ om the last term but one in Eq. [23], then replacing r_1 in those equa- ons by a_1 and averaging the $1/r_2$ in those equations over the surface of n 1, thereby replacing $1/r_2$ by $1/R$. Thus

$$\bar{\Phi}_1^\nu = \frac{e_1'}{a_1}\left(\frac{1}{D_s} - 1\right) + \frac{e_2'}{D_s R} + \frac{\nu(e_1 - e_1')}{a_1}\left(\frac{1}{D_{op}} - 1\right) + \frac{\nu(e_2 - e_2')}{D_{op}R} \quad . \qquad [25]$$

e work done during this step is W_{II},

$$W_{II} = \int_{\nu=0}^{1} \bar{\Phi}_1^\nu de_1^\nu + \int_{\nu=0}^{1} \bar{\Phi}_2^\nu de_2^\nu \quad . \qquad [26]$$

e obtains

$$I = \left(\frac{\Delta e_1 \Delta e_2}{D_{op}} - \frac{e_2'\Delta e_1 + e_1'\Delta e_2}{D_s}\right)\frac{1}{R} + \left[\frac{1}{2}(\Delta e_1)^2\left(\frac{1}{D_{op}} - 1\right) - e_1'\Delta e_1\left(\frac{1}{D_s} - 1\right)\right]\frac{1}{a_1}$$

$$+ \left[\frac{1}{2}(\Delta e_2)^2\left(\frac{1}{D_{op}} - 1\right) - e_2'\Delta e_2\left(\frac{1}{D_s} - 1\right)\right]\frac{1}{a_2} \quad . \qquad [27]$$

e net work done ΔG^r is the sum of W_I and W_{II} and is the free energy of is fluctuation. It is equal to

$$\Delta G^r = W_I + W_{II} = \left(\frac{\Delta e_1^2}{2a_1} + \frac{\Delta e_2^2}{2a_2} + \frac{\Delta e_1 \Delta e_2}{R}\right)\left(\frac{1}{D_{op}} - \frac{1}{D_s}\right) \quad , \qquad [28]$$

ere Δe_i is given by Eq. [20].

If $\Delta G_R^{0\prime}$ denotes the free energy change for unit concentration of re- tants at a separation distance R, it is related to the same quantity at finite separation $\Delta G^{0\prime}$ by an equation similar to Eq. [12], namely

$$\Delta G_R^{0\prime} = \Delta G^{0\prime} + w^p - w^r \quad . \qquad [29]$$

the intersection hypersurface of Fig. 1, the reactants and products have the same distribution of configurations (the same set values of $q^{\ddagger}$), a the same potential energy (cf Eq. [7]), averaged over a distribution of such configurations. Since the distribution of configurations (and momenta) are the same for the reactants and products on the intersection hypersurface, the entropy is also the same, and so the free energy of the reactants is the same as that for the products on the intersection hypersurface. It then follows that we can write

$$\Delta G^r - \Delta G^p = \Delta G_R^{0\prime} \quad , \qquad [3$$

where ΔG^r is given by Eq. [28] and ΔG^p by the same expression with e_i replaced by e_i^p.

To find e_1' and e_2' one minimizes the ΔG^r in Eq. [28] subject to the constraint imposed by Eq. [30].

$$\delta\Delta G^r = (\partial\Delta G^r/\partial e_1')\,\delta e_1' + (\partial\Delta G^r/\partial\Delta e_2')\,\delta e_2' = 0 \qquad [3$$

$$\delta\Delta G^r - \delta\Delta G^p = 0 \quad . \qquad [3$$

Multiplying the second equation by a Lagrange multiplier m and adding to Eq. [31], introducing expressions such as Eq. [31] for $\delta\Delta G^r$ and $\delta\Delta G^p$ into Eq. [32], and setting the coefficients of $\delta e_1'$ and $\delta e_2'$ equal to zero, one finds

$$e_1' = e_1 + m(e_i - e_i^p) \qquad (i = 1,2) \quad . \qquad [3$$

Introducing this result into Eqs. [28] and [30] one obtains

$$\Delta G^r = m^2\lambda_0 \qquad [3$$

and

$$-(2m+1)\lambda_0 = \Delta G_R^{0\prime} \quad , \qquad [3$$

where

$$\lambda_0 = (\Delta e)^2\left(\frac{1}{2a_1} + \frac{1}{2a_2} - \frac{1}{R}\right)\left(\frac{1}{D_{op}} - \frac{1}{D_s}\right) \quad , \qquad [3$$

and Δe is $e_1^p - e_1$, the charge transferred. m is the solution of Eq. [35]

The free energy barrier ΔG^* to reaction consists of the work term w^r to bring the reactants together and ΔG^r. Further, solving Eq. [35] m and introducing into Eq. [34] one obtains

$$\Delta G^* = w^r + (\lambda_0/4)\left[1 + (\Delta G_R^{0\prime}/\lambda_0)\right]^2 \quad . \qquad [3$$

In the electrode case the electrostatic potential in step 1 is giv

Eq. [14], but with the e_2 replaced by the image charge of reactant 1 the electrode $-e_1$. The image charge ensures that the potential given Eq. [14] is constant on the surface of the electrode, where $r_1 = r_2$. re is, in addition, another term due to the interaction of ion 1 with other charges on the electrode and with the surrounding electrolyte. s term is found to cancel, apart from a minor term, in steps 1 and 2 so will be omitted here for brevity. The derivation continues then before, where R now denotes the distance from ion 1 to its image, ely twice the distance to the electrode. In computing W_I and W_{II} one putes only the work to charge ion 1. One finds ultimately expressions ilar to Eqs. [34]-[37] but now we have

$$-(2m+1)\lambda_0 = -nF(E-E^{0\prime}) + w^p - w^r \qquad [38]$$

$$\lambda_0 = \frac{1}{2}\left(\frac{1}{a_1} - \frac{1}{R}\right)\left(\frac{1}{D_{op}} - \frac{1}{D_s}\right) \quad . \qquad [39]$$

t is, one obtains for this electrode case

$$\Delta G^* = w^r + (\lambda_0/4)[1 + \{-nF(E-E^{0\prime}) + w^p - w^r\}/\lambda_0]^2 \qquad [40]$$

re F is the Faraday.

This type of simplified derivation of the above results was given lier in Ref. [12], and has recently been made more readily available a review [13].

If now one wished to include simultaneously these fluctuations in vent polarization and those described earlier in reactants' vibrations, could readily do so: To the right side of Eq. [28] would be added $k_i(q_i-q_i^{0r})^2$, and $\frac{1}{2}\sum_i k_i(q_i-q_i^{0p})^2$ would be added to the correspond- expression for ΔG^p. One would then proceed as before, using Eqs. -[32], but now Eq. [31] would contain an extra term $\sum_i k_i(q_i-q_i^{0r})\delta q_i$, $\delta\Delta G^p$ would contain an analogous extra term. One would find in addi- to Eq. [33] the result that

$$q_i^{\ddagger} = q_i^{0r} + m(q_i^{0r}-q_i^{0p}) \quad . \qquad [41]$$

this $q_i^{\ddagger}$ and the e_i', again given by Eq. [33], are introduced into expression for ΔG^r, Eq. [34] would again follow, but with λ_0 replaced $_0+\lambda_i$, λ_0 being given by Eq. [36] and λ_i by Eq. [13]. Similar remarks y to the electrode reaction case.

DEDUCTIONS AND APPLICATIONS OF THE THEORY

When a statistical mechanical treatment is undertaken instead of a dielectric continuum one again obtains equations formally similar to Eq [10] to [12] and [34] to [41], but with λ_0 having a statistical mechanical value and $\lambda_i+\lambda_0$ replacing λ_i and λ_0 [3]:

$$k_r = \varkappa\ Z\ \exp(-\Delta G^*/kT) \quad , \qquad [4$$

where in the homogeneous case

$$\Delta G^* = w^r + (\lambda/4)[1+(\Delta G_R{}^{o\prime}/\lambda)]^2 \quad . \qquad [4$$

$\Delta G_R{}^{o\prime}$ is given by Eq. [29] and in the electrode case $\Delta G_R{}^{o\prime}$ is replaced

$$\Delta G_R{}^{o\prime} \rightarrow -nF(E-E^{o\prime}) + w^p - w^r \quad , \qquad [4$$

and in both cases λ is the sum of two contributions:

$$\lambda = \lambda_i + \lambda_0 \quad . \qquad [4$$

In these equations $\varkappa$, Z, w^r and w^p have the same significance as before $\Delta G^{o\prime}$ is the free energy of reaction for Eq. [4], when the reactants and products are each in unit concentration in the prevailing medium and at the prevailing temperature. $E^{o\prime}$ is the "standard" half-cell potential for Eq. [3] under the same conditions and E is the actual half-cell potential (in the absence of concentration polarization). λ for the electrode case differs from that for the homogeneous case, as before, but i each case contains a vibrational contribution λ_i from the inner coordination shell of the reactant λ_i and a contribution λ_0 from the solvent outside, as in Eq. [45]. λ_i is of the form in Eq. [13] and a dielectri continuum estimate of λ_0 is of the form in Eq. [36] or Eq. [39].

In the lecture on which the present article is based deductions fr these various equations were described, together with experimental data regarding them. Since similar material was recently presented elsewher [14], it will not be reproduced here. Instead, some of the deductions from the theoretical equations are summarized and reference is made to [14] for further details and for most of the experimental references. references both here and in [14] are intended merely to be representative, rather than complete. Several ones in addition to [14] are also cluded.

(1) Cross-Reactions and Electron-Exchange Reactions

The following reactions are known as electron exchange reactions and their rates are most frequently measured by use of isotopic tracers.

$$A_1(ox) + A_1(red) \rightarrow A_1(red) + A_1(ox) \quad [46]$$

$$A_2(ox) + A_2(red) \rightarrow A_2(red) + A_2(ox) \quad [47]$$

Let their λ's be written as λ_{11} and λ_{22}, and their rate constants as k_{11} and k_{22}. The reaction in Eq. [4] is designated as a cross-reaction with λ written as λ_{12} and rate constant as k_{12}. Because of an additivity property [3] of λ,

$$\lambda_{12} \cong \frac{1}{2}(\lambda_{11} + \lambda_{22}) \quad , \quad [48]$$

and the form of Eq. [43], one obtains a prediction of k_{12} in terms of k_{11}, k_{22} and K_{12}, the equilibrium constant of Eq. [4], when the κ's for the individual reactions are either near unity or nearly cancel, and when the work terms for the individual reactions are either small or nearly cancel:

$$k_{12} \cong (k_{11}k_{22}K_{12}f_{12})^{1/2} \quad [49]$$

where

$$f_{12} = (\ell n\, K_{12})^2/4\, \ell n(k_{11}k_{22}/Z^2) \quad . \quad [50]$$

Sutin, referenced in [14], pioneered the experimental study of these relations between cross-reactions and self-exchange reactions.

2) Dependence of the slope α of a plot of $\ell n\, k_{12}$ vs $\ell n\, K_{12}$ (Bronsted coefficient) (cf. [14]).

$$\alpha = \frac{1}{2}(1 + \Delta G_R^{0\prime}/\lambda) \quad . \quad [51]$$

3) Dependence of the slope α of a plot of ℓn i, where i is the current density for the forward reaction in Eq. [3], vs $-nF(E-E^{0\prime})$ (Tafel slope) [14].

$$\alpha = \frac{1}{2}[1 + \{-nF(E-E^{0\prime}) + w^p - w^r\}/\lambda] \quad [52]$$

4) Relations between electrode rates and homogeneous reaction rates [14].

5) Dependence of $\ell n\, k_r$ on solvent dielectric continuum properties when no specific reactant-solvent interaction exists [15].

6) Dependence of $\ell n\, k_r$ on changes in equilibrium bond lengths and angles [14].

(7) Dependence of $\ell n\ k_r$ on Δe [16].

(8) Formation of electronically-excited states, and hence chemiluminescence, in highly exothermic reactions, and properties thereof [14]

(9) Contributions to the entropy of activation of reactions [17].

(10) Electrolyte affects in electrode reactions [18] and in homogeneous reactions [17].

More recently, evidence regarding the dependence of Tafel slope in electrode reactions on $E-E^{0\prime}$ may be found in [19] and on chemiluminescence in [20]. (For further theoretical studies on references related to highly exothermic reactions, such as those involving chemiluminescen see [14].) A nice example of a test of Eq. [43] is given in [21]. An application of the present electron transfer theory to electron transfe reactions with negative activation energy [22] is found in [23]. Other interesting developments in electron transfer theory include the study reactions at semiconductor electrodes [24-26].

COMMENTS ON SEVERAL DEVELOPMENTS SINCE 1957

Elsewhere in this volume ONR Technical Report No. 12 (1957) is reproduced. It contains the formulation of this electron transfer theory for electrode reactions. That Report was concerned with the dielectric continuum contribution to the free energy barrier, a contribution for which a simplified derivation was given in an earlier section of the pre sent paper. The derivation in ONR Technical Report No. 12 is given in terms of vector-fields, vectors associated with the dielectric polarization, with the electric field due to the charges themselves, and with th total electric field due to the charges and the polarization. In the simplified derivation given in a previous section the electrostatic quan tities are not expressed in terms of vector-fields but rather in terms o scalar-fields associated with potentials and charges, though fields of a restricted kind, namely of the functional form given by Eq. [14]. A mor general derivation is given in terms of scalar-fields in [3], and vector fields in ONR Technical Report No. 12. No assumption is made initially, for example, that the reactant(s) is(are) spherical.

One generalization of the dielectric continuum result in Eqs. [36] and [39] for λ_0 is to include the fact that there is a dispersion of fre quencies characterizing the dielectric response [27,28]. The correction in the λ_0 value amounts to 18% [28].

One point of interest regarding derivation in [9,10] terms of vecto ields (viz. electric vectors) versus one in terms of scalar fields (viz. harge densities and electrostatic potentials) in [3] is the following: n the vector-field derivation in ONR Technical Report No. 12 [10] and in ts predecessor [9] the transcription of the final vector-field equations or the electrostatic free energy into scalar-field quantities was needed nd was made. This transcription is avoided by using instead scalar-ield expressions throughout.

Two of the generalizations in Ref. [3] are to use statistical mech-nics instead of dielectric continuum theory and to treat the inner shell ontribution to the fluctuations in coordinates. (This second generali-ation was given also in Ref. [29].) The generalizations led to similar redictions as those given earlier in the present paper, but without the imple dielectric equations [36] and [39] for λ_0 and hence without the imple predictions of solvent effects for reactants which do not specifi-ally interact with the solvent. A statistical mechanical expression for $_0$ was given instead, [3], one whose evaluation must await further appli-ation of the statistical mechanics of polar liquids. When a dielectric ontinuum estimate was made of this contribution one obtains the result or λ_0 given in Eq. [36] for homogeneous reactions and Eq. [39] for elec-rode reactions. When the inner contribution is included, the same final quations obtained, but with λ_0 replaced by $\lambda_0+\lambda_i$, as in Eq. [45].

In the statistical mechanical treatment it was recognized [3,29] that hen the q in Fig. 1 in the vicinity of $q^{\ddagger}$ is primarily a solvent orienta-ional or solvent vibrational coordinate the free energy change given by q. [43] is actually the free energy of a fluctuation of coordinates to alues centered on the intersection hypersurface rather than being con-ined to it. This effect introduces a minor correction factor ρ [3] in he pre-exponential factor in Eq. [42]; ρ is of the order of unity.

Among other developments since 1957 has been the recognition of the arious consequences of the theoretical equations, listed earlier. The elation given by Eq. [49] was first given essentially in 1960 [29], ap-lied to experimental data in 1963 [30], and given a more general deri-ation in 1965 [3]. The dependence of Tafel coefficient on $E-E^{0\prime}$ was ot confirmed until 1975 [19], and use was made of Mohilner's treat-ent [18] of electrolyte effects to calculate the work terms w^r and w^p. he solvent effect was tested in 1970 [15] and the effect of Δe in Eq. 39] in 1969 [16]. A comparison between some rates of electrode reactions nd the electronic structure of the reacting species, aimed at correlat-ng qualitatively with charges of equilibrium bond lengths is given in

Ref. [14]. Applications of the cross-relation Eq. [49] to biological systems have also been made and referenced in [14].

Eq. [49] has been widely tested (referenced in [14]) and has been very useful in correlating a large body of data. The main anomaly observed thus far in this equation is for reactions of Co^{+3}(aq) ion. This ion undergoes an extensive electronic rearrangement in forming Co^{+2}(aq) and a possible explanation for the anomaly is given in [14].

Quantum effects on the vibrational or solvent motion have also been treated [7,8,13,31b]. At sufficiently low temperatures the system does not use the energy ΔU^* in Fig. 1 to go from the R curve to the P curve. Instead it can "tunnel" through the barrier there. This tunneling, in many-dimensional q-space, has been treated by means of Fermi's Golden Rule for radiationless transitions and Franck-Condon factors [7,8,13,31b]. At high temperatures the expression for the free energy barrier reduces to the classical one given earlier in this paper. An analogous formalism has been used to treat [31] biological electron transfers [32] at very low temperatures.

In the quantum treatment [7,8,13] a treatment of the vibrations of the reactants is relatively straightforward, but the solvent vibrations are treated as though they are vibrations of a solid and hence undamped. In the liquid there is a strong damping. (This approximation was not made in the classical statistical mechanical treatment [3].) A modification for the damping on the quantum mechanical result has been described [33].

Again, when curve P in Fig. 1 crosses curve R on the left hand side so that the slopes of both curves are both negative at the crossing point the transition from R to P curve can only occur nonadiabatically and quantum treatments involving the Franck-Condon factors and related closely to the just mentioned quantum formalism [7,8,13], have been used, referenced in [14]. Such treatment are needed for highly exothermic reactions: When ΔU in Fig. 1 is sufficiently negative the crossing will indeed be such that the R and P curves have the same slopes at the crossing point.

Finally, there are reactions of the redox or electrochemical type where bonds are actually broken and formed during the reaction. Here, the description of the potential energy surfaces given in Fig. 1 is suggestive though not adequate. Other models have been used, the bond energy-bond order model [34] for example [35,36], and an analog to Eqs. [43], [40] and [50] has been obtained [36] but with

$$\Delta G^* = w^r + (\lambda/4) + (\Delta G_R^{0\prime}/2) + (\Delta G_R^{0\prime}/2y)\ell n \cosh y \qquad [53]$$

nd

$$f_{12} = (K_{12})^{(\ell n \cosh y)/y} \qquad [54]$$

here

$$y = (2\gamma\Delta G_R^{0\prime}/\lambda) = (\ell n\, K_{12})\gamma/\ell n(k_{11}k_{22}/Z^2)\ , \quad \gamma = \ell n\, 2\ . \qquad [55]$$

e case of the hydrogen discharge reaction, Eq. [1], will be treated in forthcoming publication [1] using a combination of nonequilibrium po- rization, BEBO, and Franck-Condon overlap methods to treat different pects of the overall problem.) Whereas the slope of a Bronsted plot omogeneous reaction) or Tafel plot (electrode reaction) was given by s. [51] and [52], they are given [36] in a BEBO model by Eq. [56] when e work terms can be neglected.

$$\alpha = \frac{1}{2}(1 + \tanh y) \qquad [56]$$

here y is given by Eq. [55]. In the electrode case the $2\Delta G_R^{0\prime}/\lambda$ is re- aced by $2\{-nF(E-E^{0\prime}) + w^p - w^r\}/\lambda$.

The theory of electron transfer reactions in solution and at elec- odes continues to be in an active and developing state. An estimate s been made for the $\varkappa$ in Eq. [42] for the ferrous-ferric electron ex- ange reaction [37]. Calculations are desirable, using increasingly ac- rate theories of electronic structure, for $\varkappa$'s of this and other reac- ons including those of $Co^{+3}(aq)$ for which extensive electronic rear- ngement occurs. Such calculations will also permit more insight into ientational effects [38] in electron transfer reactions. Again, dev- opments can be expected in statistical mechanical evaluation of the rm obtained for fluctuations in orientations of solvent molecules, eded to reach the intersection region of Fig. 1. There have been merous enlightening interactions between theory and experiment, and we n continue to look forward to this fruitful interplay in the future.

CKNOWLEDGEMENT

The author is pleased to acknowledge the support of this work by the fice of Naval Research.

REFERENCES

1 R. A. Marcus, Proc. Int. Conf. Phys. Chem. and Hydrodynamics, Hemisphere Publ. Corp., Washington, D. C., 1977 (to be submitted).

2 cf R. A. Marcus, Ann. Rev. Phys. Chem., 15(1964)155.

3 R. A. Marcus, J. Chem. Phys., 43(1965)679. In this paper the reaction rate is calculated instead for the reverse step in Eq. [3], whil in Ref. [10], as in the present paper, the forward step in Eq. [3] discussed.

4 R. R. Dogonadze and Yu. A. Chizmadzhev, Proc. Acad. Sci. U.S.S.R., Phys. Chem. Sec., English Transl., 145(1962)563.

5 L. Landau, Physik Z. Sowjetunion, 1(1932)88; 2(1932)46; C. Zener, Proc. Roy Soc. (London), A137(1932)696; A140(1933)660; E. C. G. Stueckelberg, Helv. Phys. Acta, 5(1932)369.

6 E.g., J. S. Cohen, S. A. Evans and N. F. Lane, Phys. Rev., A4(1971) 2248; V. Babamov, Ph.D. Thesis, University of Illinois, 1977 (to be submitted) (review and numerical methods).

7 V. G. Levich, in H. Eyring, D. Henderson and W. Jost (Editors), Physical Chemistry, An Advanced Treatise, Vol. 9B, Academic Press, New York, 1970.

8 R. R. Dogonadze, in N. S. Hush (Editor), Reactions of Molecules at Electrodes, Wiley, New York, 1971, p. 135.

9 R. A. Marcus, J. Chem. Phys., 24(1956)966.

10 R. A. Marcus, ONR Technical Report No. 12, Project NR 051-331, 1957

11 P. Lorrain and D. R. Corson, Electromagnetic Fields and Waves, Freeman and Co., San Francisco, 1970, 2nd Ed., Eq. [2-92] and p. 68.

12 R. A. Marcus, in P. Kirkov (Editor), Lecture Notes, International Summer School on the Quantum Mechanical Aspects of Electrochemistry Ohrid, Yugoslavia, 1971.

13 P. P. Schmidt, Electrochemistry Spec. Period. Rep. Electrochem., 5(1975)21. This article by Schmidt is an excellent review of the theoretical literature.

14 R. A. Marcus, in E. D. Goldberg (Editor), Dahlem Konferenz on the Nature of Sea Water, Phys. Chem. Sci. Res. Rept. 1, Abakon Verlag, Berlin, 1975, p. 477.

15 J. R. Brandon and L. M. Dorfman, J. Chem. Phys., 53(1970)3849.

16 M. E. Peover and J. S. Powell, J. Electroanal. Chem., 20(1969)427.

17 E. Waisman, G. Worry and R. A. Marcus, J. Electroanal. Chem. Interfacial Electrochem. (in press).

18 D. M. Mohilner, J. Phys. Chem., 73(1969)2652.
19 P. Bindra, A. P. Brown, M. Fleischmann and D. Pletcher, Electroanalyt. Chem. Interfacial Electrochem., 58(1975)39; cf F. C. Anson, N. Rathjen and R. D. Frisbee, J. Electrochem. Soc., 117(1970)477.
20 N. Periasamy, Ph.D. Thesis, University of Bombay, Tata Institute of Fundamental Research, Bombay, August 1976; N. Periasamy and K. S. V. Santhanam, Chem. Phys. Letters, 39(1976)265.
21 D. Meisel, Chem. Phys. Letters, 34(1975)263.
22 J. N. Braddock, J. L. Cramer and T. J. Meyer, J. Amer. Chem. Soc., 97(1975)1972.
23 R. A. Marcus and N. Sutin, Inorg. Chem., 14(1975)213.
24 J. F. Dewald, in N. B. Hannay (Editor), Semiconductors, ACS Monograph, Reinhold, New York, 1959, p. 727.
25 H. Gerischer, in H. Eyring (Editor), Physical Chemistry, An Advanced Treatise, Vol. 9A, Electrochemistry, Academic Press, New York, 1970, p. 463.
26 R. Memming and F. Möllers, Ber. Bunsenges. Phys. Chem., 76(1972)475; M. Gleria and R. Memming, Z. Phys. Chem., (Frankfurt am Main) 101 (1976)171.
27 A. A. Ovchinnikov and M. Ya Ovchinnikova, Soviet Phys. JETP, English Trans., 29(1969)688; Soviet Physics Dokl., English Transl., 14(1969) 447.
28 R. R. Dogonadze, Ber. Bunsenges. Phys. Chem., 75(1971)628.
29 R. A. Marcus, Discussions Faraday Soc., 29(1960)21.
30 R. A. Marcus, J. Phys. Chem., 67(1963)853,2889.
31 (a) J. J. Hopfield, Proc. Natl. Acad. Sci. U.S.A., 71(1974)3640;
(b) J. Jortner, J. Chem. Phys., 64(1976)4860.
32 D. DeVault, J. H. Parker and B. Chance, Nature (London), 215(1967)642; D. DeVault and B. Chance, Biophys. J., 6(1966)825.
33 P. P. Schmidt, J. Chem. Soc. Faraday Trans. 2, 69(1973)1122.
34 H. S. Johnston, Adv. Chem. Phys., 3(1960)131.
35 M. Salomon, C. G. Enke and B. E. Conway, J. Chem. Phys., 43(1965)3989.
36 R. A. Marcus, J. Phys. Chem., 72(1968)891; Faraday Symp. Chem. Soc., 10(1975)60.
37 J. A. Jafri, G. Worry, M. D. Newton, N. Sutin and R. A. Marcus, paper presented at 173rd National A.C.S. Meeting, New Orleans, La., March, 1977.
38 R. Chang and S. I. Weissman, J. Amer. Chem. Soc., 89 (1967)5968.

OFFICE OF NAVAL RESEARCH
Contract Nonr 839(09)
Task No. NR 051-339
TECHNICAL REPORT NO. 12

On the Theory of Overvoltage for Electrode Processes Possessing Electron Transfer Mechanisms. I.

by
R. A. Marcus

Foreword (March 1, 1977)

The following article was first prepared as the above Office of Naval Research Technical Report, and is being reproduced here in its entirety. The results were applied in subsequent papers on electrode kinetics [1], and the Report itself has been widely cited in the literature.

Originally intended for journal publication, the article was never submitted because of the efforts of the author towards generalization. Th generalization was published some years later [2], but in some respects the approach of the 1957 Technical Report is more readable. The princip generalizations are the inclusion of inner coordination shell contributio and the use of statistical mechanics for inner and outer contributions [2,3

The 1957 Report and a predecessor [4] had as their aims the treatmen of the outer shell contribution; their equation for the dielectric continuum calculation of that contribution is the same as that in the later publication [2]. The Report draws upon some derivations in ONR Technica Report No. 11, and this latter Report is included also. The generalization of those results may be found in Ref. [5].

Elsewhere in this book we comment on more recent developments.

REFERENCES

1 R. A. Marcus, Can. J. Chem., 37(1959)138; R. A. Marcus, in E. Yeager (Editor), Transactions Symposium Electrode Processes, 1959, John Wiley, New York, 1961, p. 239.
2 R. A. Marcus, J. Chem. Phys., 43(1956)679.
3 R. A. Marcus, Discuss. Faraday Soc., 29(1960)21; cf R. A. Marcus, Trans. N. Y. Acad. Sci., 19(1957)423 for a result on the inner shell contribution.
4 R. A. Marcus, J. Chem. Phys., 24(1956)966.
5 R. A. Marcus, J. Chem. Phys., 38(1963)1335,1858; ibid., 39(1963)460,173

THE THEORY OF OVERVOLTAGE FOR ELECTRODE PROCESSES POSSESSING ELECTRON ANSFER MECHANISM. I.*

A. MARCUS
partment of Chemistry, Polytechnic Institute of Brooklyn, Brooklyn, N.Y.

STRACT

A quantitative theory of overvoltage is developed for electrode ocesses possessing electron transfer mechanisms. The assumptions and neral approach of the theory are related to those used in the writer's cent work on homogeneous oxidation reactions involving electron transrs.

A simple expression is derived for the electron transfer rate nstant as a function of the activation overvoltage and other quanties. The latter include the ionic radii and charges, and the work reired to transport the ion to the electrode surface. From this exession for the rate constant, deductions can be made concerning salt fects, temperature coefficients, transfer coefficients, and the relaon between exchange currents and the rates of isotopic exchange retions in solution. Some recent data are briefly discussed from this int of view and some suggestions made for further studies.

The theory and the final equation are free from arbitrary assumpons and adjustable parameters. Several possible refinements are noted.

TRODUCTION

The detailed mechanisms of many electrode processes have been udied in recent years using A.C. and D.C. methods [1]. Some of these chanisms consist only of a simple electron transfer, while in others is step is preceded or followed by chemical equilibria and by other emical reactions.

From such studies, similarities have been noted between the rate th which certain species are oxidized or reduced by electrochemical d by chemical means. Various authors [2] have interpreted the rates

---

ffice of Naval Research, Technical Report No. 12, Contract Nonr 839(09), sk No. NR 051-339.

of electrochemical electron transfers in terms of concepts related to those used for electron transfers in chemical reactions. A brief qualitative comparison of the relative rates of several isotopic exchange redox reactions and of the corresponding electrochemical systems has been given by the writer [3].

Recently, a quantitative theory has been developed for calculating the rate constant of redox reactions involving an electron transfer in solution [4,5]. It proceeded from first principles plus assumptions which appear reasonable on a priori grounds. The general approach of the theory and the final theoretical equations were free from arbitrary assumptions and adjustable parameters. An analogous theory will be formulated here for electron transfer electrode processes. As before, the theory is not applicable to atom transfer mechanisms (hydrogen overvoltage, for example).

The Electron Transfer Rate Constant

Equations describing any overall electrochemical process may include terms for the transport of ions to the electrical double layer region at the electrode, for any chemical reactions of the electrochemically active species, for the work required to penetrate the double layer (if necessary), and for the actual electron transfer. Only under certain conditions can these factors be disentangled in a relatively simple way and simple electron transfer rate constants defined: The double layer region should be sufficiently thin that (1) no chemical reaction occurs in it and (2) diffusion across it is sufficiently rapid so that the concentration of an ion at any point in the double layer is related to its concentration just outside it by the work required to transport it to that point. At appreciable salt concentrations the double layer thickness appears to be only of the order of several Angstroms [6].

Under the above conditions, one can describe the entire electrochemical process by force-free differential equations [1] (containing any chemical reaction terms [7], if necessary) outside of the double layer region, while the boundary condition, at a boundary surface S drawn just outside the double layer, contains rate constants which depend only on the electron transfer process itself. This boundary condition is that the flux of any ion through this surface S equals its net rate of disappearance by electron transfer.

If A and B represent the electrochemically active ion before and fter its electron transfer with the electrode, the electron transfer tep may be written as

$$A + ne \underset{k_b}{\overset{k_f}{\rightleftharpoons}} B, \qquad [1]$$

here n is the number of electrons lost by the electrode. If c_A^S and c_B^S enote the concentrations of A and B just outside the double layer, then he electron transfer constants, k_f and k_b, will be defined by the elation

$$\text{Net electron transfer rate per unit area} = k_f\, c_A^S - k_b\, c_B^S. \qquad [2]$$

ote: To avoid confusion with references, equations are referred to by) in the body of the text.]

EORY

asic Assumptions

The basic assumptions employed here will be analogous to those made n the oxidation-reduction reaction theory [4].

(a) In the activated state of reaction (1) the spatial overlap be- ween the electronic orbitals of the two reactants is assumed to be small. n the present instance the "reactants" are the electrode and the dis- harging ion or molecule (to be referred to as the "central ion"). This ssumption has been discussed previously [4] and some preliminary evi- ence in favor of it was cited.

(b) The central ion is treated as a sphere within which no changes n interatomic distances occur during reaction (1) and outside of which he solvent is treated as a dielectrically unsaturated continuum. Some scussion of this assumption has been given [4]. For complex ions or ydrated monatomic cations the sphere includes the first coordination hell of the central atom [4,5a]. A preliminary suggestion has been ade for the effective size of unsymmetrical organic molecules and has een discussed [5b]. A refinement of the theory which takes changes, if y, in interatomic distances into consideration will be presented later.

ture of the Activated Complex

It follows [4] quantum mechanically from the first assumption that successful electron transfer between the central ion and the electrode oceeds via two successive intermediate states, X^* and X, say. These ates have the same atomic configuration and the same total energy, but X^* the electronic configuration of the central ion is that of A and X, it is that of B. The electronic configuration of the electrode undergoes a corresponding change, deduced later [8].

By arguments similar to those employed in the redox theory [4], it then follows that the actual electron transfer (the formation of X from X^*) must be preceded by a reorientation of the solvent molecules in the vicinity of the discharging ion and nearby area of the electrode. For similar reasons, a change in ionic atmosphere in this region also precede the electron transfer. The new configuration of the solvent and ionic atmosphere, which is the same in X^* and in X, will prove to be intermedi between that in the initial state and that in the final state. As such, it is not that which is predicted by the charge distribution in X^* or by that in X. That is, it is not in electrostatic equilibrium with either charge distribution, and its properties cannot be described by the usual electrostatic expressions. Instead, expressions which take this nonequil brium behavior into account must be used. Recently, such equations have been derived for electrode systems [9].

Once again [4], there are an infinite number of pairs of states, X^* and X, satisfying the constant energy-atomic configuration restriction, and we shall be interested in determining the properties of the most probable pair, i.e. the one with the minimum free energy of formation fr the initial state. This is done by minimizing the expression for this free energy of formation subject to the restriction that X^* and X have t same atomic configuration and the same energy.

Reaction Scheme

From the preceding discussion we may write for the mechanism of reaction [1] the following scheme, analogous to that in Reference 4.

$$A \underset{k_{-1}}{\overset{k_1}{\rightleftharpoons}} X^*$$

$$X^* \underset{k_{-2}}{\overset{k_2}{\rightleftharpoons}} X$$

$$X \xrightarrow{k_3} B.$$

Here, a central ion just outside the double layer is denoted in its initial electronic state by A and in its final state by B. As noted earlier, step (3) involves a suitable reorganization of configuration of the solvent and ionic atmosphere, and (if necessary) a suitable

enetration of the electrical double layer. Step (4) is the actual elec-
ron transfer itself, and step (5) involves a reversion of configuration
of solvent and atmosphere) to one in equilibrium with the new charge on
ne central ion. It also involves a motion away from the electrode. As
n the redox theory, the reverse of (5) occurs but it need not be con-
idered in the computation of k_f.

Steady-state considerations for c_X* and c_X lead to the relation [4]

$$k_f = k_1 /[1 + (1 + k_{-2}/k_3)\ k_{-1}/k_2]. \qquad [6]$$

As discussed later, when the probability of electron transfer in the
fetime of the intermediate state X^* (10^{-13} sec) is large, k_f is about
lf of k_1. k_1 depends on the free energy of formation of X^* from the
itial state in reaction (1). We proceed first to the calculation of this
ee energy change and later to a discussion of the evaluation of the
te constants. With this aim in view, we consider in the next section
e electrochemical system, the question of equilibrium (electrostatic
d electrochemical), and fluctuations therefrom.

ectrostatic vs Electrochemical Equilibrium

The electrochemical cell in which the reaction under investigation
curs may be considered to consist of two half-cells, joined with or
thout a salt bridge. One of the half-cells will be taken to be re-
rsible, and only at this electrode will there be electrochemical equi-
brium. At the other electrode, M say, where reaction (1) occurs, there
ll nevertheless be electrostatic equilibrium because of the small re-
xation time of the ionic double layer [10]. The formation of X^*
presents a fluctuation from this equilibrium. In the following pages,
shall, for brevity, denote "electrostatic equilibrium" by "equilibrium."
en "electrochemical equilibrium" is intended, it will be explicitly
ated.

When an electron transfer occurs at M, completion of the electro-
emical process requires the transfer of an equivalent number of elec-
ons at the reversible electrode and of an equivalent number of ions
om one half-cell to the other. This other transfer actually represents
fluctuation from electrochemical equilibrium and accordingly involves
free energy change. The fluctuation may occur before, after or during
action (1), but k_f is the same in each case. For convenience of
nputation, we can let it occur before or after, so that the initial,

intermediate, and final states of the system in the reaction sequence (3 to (5), designated by A, X^*, X and B, have the same number of each type of ionic species, except for the particular central ion undergoing electron transfer.

Free Energy of Formation of State X^*

The electrostatic contribution, ΔF^*, to the free energy of formatio of state X^* from state A will be different from zero because of (1) the work which may be required to transport the central ion from a point jus outside the double layer to some particular point in it (if necessary), and (2) the work required to reorient the solvent molecules and the ioni atmosphere to a nonequilibrium configuration, for this position of the central ion.

Let F_e^A, F_e^*, F_e, and F_e^B denote the electrostatic free energy of the electrode M and of its solution when the system is in the state A, X^*, X and B, respectively. (Throughout, asterisks will be used to designate the properties of X^* and will be omitted in designating the properties of X.) We have therefore for ΔF^*,

$$\Delta F^* = F_e^* - F_e^A \qquad [7$$

F_e^* is given by the general expression (8) and F_e is given by a similar equation, minus the asterisks [9a].

$$F_e^* = \frac{1}{2}\int[\underset{\sim}{E}_v{}^* \cdot \underset{\sim}{E}_v{}^* - \alpha_e \underset{\sim}{E}^* \cdot \underset{\sim}{E}_v{}^* - \underset{\sim}{P}_u \cdot (\underset{\sim}{E}^* + \underset{\sim}{E}_v{}^* - \underset{\sim}{P}_u/\alpha_u)]\ dV + kT\sum_i \int c_i \ln c_i/c_i{}^o\ dV + \chi\int_M \sigma^*\ dS \qquad [8$$

where $\underset{\sim}{E}_v{}^*(\underset{\sim}{r})$ = electric field at point r exerted in a vacuum by all the ionic charges in the state X^* and by those electrode charges which they would induce in a vacuum.

$\underset{\sim}{E}^*(\underset{\sim}{r})$ = electric field in state X^*. It equals $-\nabla\varphi^*$.

$\varphi^*(\underset{\sim}{r})$ = inner potential in state X^*. (cf Eq. (5) of Reference 9a)

α_e = E-type[9] polarizability = $(D_{op}-1)/4\pi$. [9

α_u = U-type[9] polarizability = $(D_s - D_{op})/4\pi$. [10

$c_i(\underset{\sim}{r})$ = concentration of ions of type i in states X^* and X.

$c_i{}^o$ = average concentration of ions of type i. It equals thei number n_i in the solution divided by the latter's volume

$\underset{\sim}{P}_u(\underset{\sim}{r})$ = U-type polarization in states X^* and X.

χ = potential drop at electrode-solution interface due to an oriented solvent dipolar layer.

The volume integrals in this equation are over the volume V, and the surface integral is over the surface of electrode M. In Eqs. (9) and (10) D_S and D_{op} are the static dielectric constant and the square of the refractive index, respectively. The functions $c_i(\underset{\sim}{r})$ and $\underset{\sim}{P}_u(\underset{\sim}{r})$ in state X* are the same as those in state X because of the constant atomic configuration restriction.

F_e^A is given by the same type of equation as (8), but with the equilibrium relations for $\underset{\sim}{P}_u(\underset{\sim}{r})$ and $c_i(\underset{\sim}{r})$ introduced. (cf Eqs. (21) and (22) of Reference 9a). F_e^B is given by a similar equation, but the functions in the integrand now refer to state B. Eq. (8) and subsequent equations treat X as being independent of the average electrode charge density. However, in Appendix X it is shown that the final equations, Eqs. (25) to (27), are unchanged even if X were a function of this quantity.

Restraint Imposed by the Constant Energy - Atomic Configuration Restriction

The free energy of formation of X^* from state A consists of the electrostatic contribution ΔF* and of a term, described later, associated with the localization of the center of gravity of the central ion in a narrow region near the electrode [11]. The free energy of formation of X from B contains an electrostatic term $F_e - F_e^B$, ΔF say, and a center of gravity term equal to that noted previously.

As in the redox theory, it follows from assumption (a) given earlier that no energy change and no configurational entropy change accompany the formation of state X from X^*. The electronic entropy change arising from any possible change in the electronic degeneracy of the central ion and of the electrode is zero or negligible. Accordingly, X* and X have the same free energy.

The net free energy change in forming B from A is therefore

$$F_e^* - F_e^A + (F_e^B - F_e) = \Delta F^* - \Delta F \qquad [11]$$

Independently, this free energy of reaction (1) can be written as the sum of the following terms:

(a) The change in chemical potentials of ions A and B and of the electron in electrode M. This is $\mu_B - \mu_A - n\,\mu_e^M$, where n is the number of electrons transferred in reaction (4).

(b) the change in free energy due to the transfer of a charge $(e^* - e)$ from a solution of inner potential φ_S to a metal of of inner potential φ_M, e* and e denoting the charges of ions A and B, respectively. This term is $(e^* - e)(\varphi_M - \varphi_S)$, which we shall denote by $(e^*-e)\,\Delta\varphi$.

If electrode M were in electrochemical equilibrium with the actual concentrations of A and B just outside the double layer, then the sum of these two terms would be zero. The corresponding value of $\Delta\varphi$, denoted by $\Delta\varphi'$, equals the chemical potential term in (a) divided by $-(e^* - e)$.

Using this relation, the net free energy change, (a) plus (b), then becomes:

$$(e^*-e)(\Delta\varphi - \Delta\varphi') = (e^* - e)\,\eta_a, \qquad [12$$

where η_a is the activation overvoltage defined by Eq. (12).

Equating this to expression (11), we obtain the equation of restrain imposed by the constant energy - constant atomic configuration restrictio

$$\Delta F^* - \Delta F = (e^* - e)\,\eta_a \qquad [13$$

This equation is the equivalent of Eq. (19) of the redox theory [4].

It may be remarked that $(\mu_B-\mu_A)$ is related to its standard value, $(\mu_B^o-\mu_A^o)$, and $\Delta\varphi$ to its standard value, $\Delta\varphi^o$, by Eqs. (14) and (15), where the f's denote activity coefficients and the c^S's denote concentrations just outside the double layer.

$$\mu_B - \mu_A = \mu_B^o - \mu_A^o + kT\ \ln\,(f_B/f_A)(c_B^S/c_A^S) \qquad [14]$$

$$(e-e^*)(\Delta\varphi - \Delta\varphi^o) = kT\ \ln\,(f_B/f_A)(c_B^S/c_A^S) \qquad [15]$$

Minimization of ΔF^* Subject to Restraint Imposed by Eq. (12)

We minimize ΔF^* with respect to arbitrary variations $\delta P_u(r)$ and $\delta c_i(r)$.

$$\delta\Delta F^* = 0 \qquad [16]$$

This variation is performed subject to the condition of fixed number of ions n_i and subject to Eq. (12) at fixed η_a.

$$\delta n_i = \int \delta c_i\ dV = 0 \qquad [17]$$

$$\delta\Delta F^* - \delta\Delta F = 0 \qquad [18]$$

Introducing Eqs. (7) and (11) for ΔF^* and ΔF into Eqs. (16) and (18), we note that δF_e^A and δF_e^B equal zero since they are independent of δP_u and δc_i, while δF_e^* and δF_e are given by Eq. (20) of Reference 9a. Introducing these results into Eqs. (16) to (18), we obtain

$$-\int (\mathbf{E}^* - \mathbf{P}_u/\alpha_u)\cdot \delta\mathbf{P}_u \, dV + \sum_i \int (kT \ln c_i/c_i^o + e_i \varphi^*) \, \delta c_i \, dV = 0 \quad [19]$$

$$\int (\mathbf{E} - \mathbf{E}^*)\cdot \delta\mathbf{P}_u \, dV + \sum_i e_i \, (\varphi^* - \varphi) \, \delta c_i \, dV = 0 \quad [20]$$

Multiplying Eq. (20) by a constant m, the Lagrangian multiplier, multiplying Eq. (17) by another multiplier, $-\ln 1_i$ say, and adding to Eq. (19), we obtain

$$\int(-\mathbf{E}^* - m\,\mathbf{E}^* + m\,\mathbf{E} + \mathbf{P}_u/\alpha_u)\cdot \delta\mathbf{P}_u \, dV + \sum_i \int [kT \ln c_i/c_i^o + e_i(\varphi^* + m\varphi^* - m\varphi) - \ln 1_i]\,\delta c_i \, dV = 0 \quad [21]$$

This equation is an identity for all arbitrary variations $\delta\mathbf{P}_u(\mathbf{r})$ and $\delta c_i(\mathbf{r})$, so that the cœfficients of these quantities equal zero at each point $\mathbf{r}$.

$$\mathbf{P}_u = \alpha_u \, (\mathbf{E}^* + m\,\mathbf{E}^* - m\,\mathbf{E}) \quad [22]$$

$$c_i = c_i^o \, 1_i \, \exp\,[-e_i(\varphi^* - m\varphi^* - m\varphi)/kT] \quad [23]$$

By integrating both sides ofEq. (23) over the volume V and setting the LHS equal to n_i, 1_i can be determined.

$$1_i = V/\int \exp\,[-e_i(\varphi^* + m\varphi^* - m\varphi)/kT] \, dV \quad [24]$$

Eqs. (22) to (24) describe the solvent and ionic configuration in the pair of intermediate states X* and X.

Equation for ΔF^*

In order to compute ΔF^* in the most direct manner, Eqs. (22) and (23) for $\mathbf{P}_u$ and c_i are first introduced into expressions for $\varphi(\mathbf{r})$ and $\varphi^*(\mathbf{r})$, cf Eq. (5) of Reference 9a). The resulting integral equations for these potentials are then solved and $\mathbf{E}$ and $\mathbf{E}^*$ computed from them. F_e^* is then calculated from Eq. (8), F_e is obtained from an analogous equation, and an expression for ΔF^* is then derived.

However, we shall find it convenient to proceed more indirectly in the following manner which leaves the problem of solving the actual integral equations for the last step.

We first show (Appendix I) that there is a hypothetical equilibrium system (i.e. equilibrium configuration of solvent and ionic atmosphere) having the same $\mathbf{P}_u(\mathbf{r})$ and $c_i(\mathbf{r})$ as states X* and X, but having different charges on the central ion and on the electrode. The various properties of this hypothetical system are determined in Appendix I. F_e^* is next expressed in terms of the electrostatic free energy of the hypothetical system, $F_e^\dagger$, in Appendix III. In Appendix IV, $F_e^\dagger$ is then related to F_e^A and F_e^* F_e. After obtaining expressions in Appendix V for $\underset{\sim}{E}*$ and $\underset{\sim}{E}$, $F_e^* - F_e$ is evaluated in Appendix VI. The final theoretical equations for ΔF^*, Eqs. (25) to (27) below, are then deduced from this value of $F_e^* - F_e$ in Appendices VI and VII, after utilizing some relations obtained in Appendix VIII.

In this way it is shown that

$$\Delta F^* = w^* + \frac{m^2}{2}\lambda \qquad [25]$$

where m satisfies the equation

$$-(m + \frac{1}{2})\lambda = (e^* - e)\eta_a + w - w^* \qquad [26]$$

and where

$$\lambda = (\Delta e)^2 \left(\frac{1}{a} - \frac{1}{R}\right)\left(\frac{1}{D_{op}} - \frac{1}{D_s}\right) \qquad [27]$$

In these equations, Δe denotes $(e^* - e)$, the charge transferred to the electrode in reaction (1); <u>a</u> denotes the effective radius of the central ion, discussed earlier; R/2 is the distance of the ion to the electrode surface in the state X^*; w^* and w denote the work required to transport the central ion from the body of the solution to a distance R/2 from the electrode surface when the ion is in its initial state, respectively; the remaining quantities have been defined previously.

We shall return later to a discussion of the work terms w^* and w in Eqs. (25) and (26).

Rate Constants of Elementary Steps

(a) Estimation of k_{-1} and k_3

The rate constants k_{-1} and k_3 in reactions (3) and (5) are associate with the disappearance of X^* and of X, as a result of disorganization of the solvent polarization or of a suitable motion of the central ion [4]. As shown previously [4], these constants are equal to each other and have a value of about 10^{13} sec^{-1}.

(b) Estimation of k_2 and k_{-2}

The ratio k_2/k_{-2} is the equilibrium constant of reaction (4), exp

$(\Delta S_e/R)$, ΔS_e being the electronic entropy change accompanying this reaction. As noted earlier, this change is either zero or negligible, and we have $k_2 \sim k_{-2}$. Eq. (6) thus reduces to

$$k_f = p\ k_1 \qquad [28]$$

where p, a function of R defined by Eq. (29), is the probability that a successful electron transfer occurs during the lifetime of the intermediate state, 10^{-13} sec.

$$p(R) = 1/(2 + k_3/k_2) \qquad [29]$$

When the overlap between the electronic orbitals of the ion and electrode in the intermediate state is not too small, then k_2 will be of the same order or larger than k_3 and we will have

$$p \cong \frac{1}{2}\ ,\ k_f \cong k_1/2 \qquad [30]$$

The evaluation of k_2 and p(R) for electron transfers between ions in solution was discussed previously, several methods being available [4]. A preliminary calculation for one system indicated that p was of the order of magnitude of unity [12], but a more detailed estimate would be desirable. In preliminary applications, we shall take $p \sim 1/2$. Its value in various systems can be inferred from the temperature-independent factor in the experimental value of k_f, when the work terms w^* and w are small. If, because of any large repulsion between ion and electrode, R is especially large, then p will be correspondingly small.

(c) Evaluation of k_f

We first observe that k_f in Eq. (28) can be expressed in terms of the equilibrium constant for the formation of X^* in reaction (3).

$$k_f = p\ (k_1/k_{-1})\ k_{-1} \qquad [31]$$

If the reaction were confined to a plane, the value of k_f could be shown to be Z_w, the collison frequency of an uncharged molecule with unit area of the electrode, multiplied by the value of $p \exp(-\Delta F^*/kT)$ in that plane [12a]. Z_w equals $\sqrt{kT/2\pi m}$, which is about 10^4 cm sec^{-1} at 25°C, for an ion of mass 100 gm mole^{-1}.

Unlike atom transfer processes, however, electron transfers at electrodes can occur at various R's, rather than just at the plane of the electrode-solution interface. Taking this fact into consideration, Eq. (32) can be deduced [13].

$$k_f \cong 5 \times 10^4 \; p \; e^{-\Delta F^*/kT} \text{ cm sec}^{-1} \qquad [32]$$

Of the various terms appearing in Eq. (25), the values of the radii a have been discussed elsewhere [4,5]. The value of R is that which maximizes p exp(-$\Delta F^*/kT$). When w^* and w are zero (sufficiently high salt concentrations) or when they do not tend to increase ΔF^*, R equals its minimum value. This is 2a, if an oriented solvent layer at electrode-solution interface does not hinder the approach of the ion. (cf Appendix X.)

The term w^* can be evaluated by solving the usual Poisson-Boltzmann equation when the central ion A is at a distance R/2 from the electrode, solving it when the ion is in the body of the solution, and then introducing these solutions for the electrostatic potential into the appropriate equations for the electrostatic free energy of the system (Eqs. (39) and (41) of Appendix II). The difference in electrostatic free energy is w^*. Similarly, w can be computed from analogous equations for the central ion B.

However, a considerably simpler though less rigorous procedure has generally been assumed for calculating the work required to transport an ion to some distance (R/2) from the electrode. The assumption is generally made that this work equals the charge on the central ion multiplied by the difference in the value of the electrostatic potential at R/2 and in the body of the solution (i.e., outside the double layer), this potential being computed in the absence of the central ion. This particular potential can also be inferred from measurements on the electrical double layer by various methods [6,14]. In this way Franklin [14] has obtained a correlation between salt effects on hydrogen overvoltage, double layer computations, and measurements on the electrical double layer.

APPLICATIONS OF THE THEORETICAL EQUATION

A detailed quantitative application of Eq. (32) for k_f, using Refinement (1) noted in the following section when necessary, will be presented in a later paper. However, several preliminary remarks will be made here:

(1) Little information is available on salt effects for simple electron transfer systems. It would be particularly desirable to see

what correlations exist between k_f values (at $\eta_a = 0$) and double layer measurements in the presence of various salts. Judging from data on isotopic exchange reactions, the use of electrochemically active ions which do not readily form specific complexes with the salt, such as the cobalt ethylenediamines, the cobalt amines, the manganate-permanganate ions and other complex ions, e.g. the iron cyanides perhaps, rather than hydrated metal cations, would be of especial interest.

(2) When the temperature-dependent work terms w* and w are zero sufficient concentration of added salt), the temperature-independent factor equals 5×10^4 p cm sec^{-1}. Randles and Somerton [15] found values or this factor usually in the range $10^3 - 10^5$ cm sec^{-1}, suggesting perhaps that p has its maximum value, 1/2. However, further studies on temperature-independent factors using complex ions, such as those indicated earlier, and on the effects of added salt would be of particular interest.

(3) When the work terms w* and w are zero, the transfer coefficient defined as $-[\partial \ln i/\partial(\eta_a \Delta e/kT)]_T$, where i is the electrical current density in the forward direction) is 0.50. Using platinum electrodes, Rubin and Collins [16] found a value of about 0.42 for the $Fe(CN)_6^{-4}$ system and .46 for the $Mo(CN)_8^{-4}$ - $Mo(CN)_8^{-3}$ couple when the concentration of added potassium chloride was 0.1 M.

(4) The theoretical equation bears a very close relationship to that obtained for isotopic exchange reactions [4], in qualitative agreement with the data [3]. This close relationship is preserved in the Refinement (1) discussed in the following section. In a subsequent paper it is planned to discuss this correlation in a quantitative way. However, more detailed experimental studies on the rates of simple electron transfer electrode processes whose isotopic exchange counterparts have been investigated would be especially interesting.

The general sparsity of data on electrode kinetics of <u>simple</u> electron transfers can be ascribed to experimental difficulties in measuring the very fast rates of these systems. Recent innovations in technique and interpretation, such as those introduced in Reference 15 and those summarized in Reference 1, have begun to remedy this situation.

REFINEMENTS

Some refinements of the theory described in this paper are noted below.

(1) Changes in Interatomic Distances within the Central Ion

When a particular interatomic distance within the central ion is different in the two valence states, A and B, a change in this distance will occur during reaction (1) and this will contribute to ΔF^*. The effect appears [5a] to be appreciable for hydrated metal cations such as Fe^{+2} - Fe^{+3} and small for others, such as MnO_4^- - MnO_4^{-2}, $Fe(CN)_6^{-3}$ - $Fe(CN)_6^{-4}$. In unpublished work (cf [3]) we have computed it for the electron transfers between ions in solution in terms of certain force constants and of differences in interatomic distances. A similar calculation will be described for electrode processes in a later paper.

(2) Quantum Limitation of the Image Force Theory

The electrical image force theory was employed in the derivation of Eq. (25) for ΔF^*. Recently the quantum limitation on the image theory was estimated by Sachs and Dexter [17] and their results can be applied here. We first note that ΔF^* equals its value at $R = \infty$ minus an amount whose maximum value, which occurs at $R = 2a$, is half of this R_∞ value when the work terms are zero. The quantum limitation causes no error in the R_∞ value [17], while the error in the second term is perhaps about 8 when the distance from the ion to the electrode is 5 Å [18], so that the overall error is about 8% or less.

(3) Electrode Double Layer Containing Specifically Adsorbed Ions

Until now, we have omitted from consideration electrical double layers in which some of the ions are specifically adsorbed on the electrode. In the case of specifically adsorbed ions which do not form specific complexes with the central ion itself, a rigorous analysis wou take into account the detailed manner in which the adsorbed ionic layer was perturbed by the approach of the central ion. In the interest of simplicity, a less accurate analysis will be undertaken here.

Let us consider the simplest case - one in which the adsorbed ions are essentially not perturbed by the approach of the central ion. (Thi assumption is indeed implied in current interpretations of measurements of salt effects on hydrogen overvoltage [14].) We then may treat the adsorbed ions as fixed in position. Let us regard each of these ions as being a sphere having a surface density equal to its charge divided by its area [19].

Under this condition, Eq. (8) still applies for the value of F_e^*. Since the positions of these ions are fixed, they cannot vary during th inimization process in Eqs. (16) to (18). The c_i there refer to the ions ı the diffuse portion of the electrical double layer. As a result, it s found that the subsequent equations, Eqs. (22) and (23) are still ap- licable. It is shown in Appendix IX that the final equations for ΔF^* lso remain intact. The terms w and w* now represent the work required ɔ transport an ion of charge e and e*, respectively, in the presence of ıe adsorbed layer.

CKNOWLEDGEMENT

The writer would like to acknowledge the support of this research by ıe Office of Naval Research under Contract Nonr 839(09), and by the ational Science Foundation.

PPENDIX I. PROPERTIES OF AN EQUILIBRIUM SYSTEM HAVING THE SAME $c_i(\tilde{r})$ AND $\tilde{P}_u(\tilde{r})$ AS X*

The properties of an equilibrium system will be denoted in this Ap- endix by the superscript †.

At any point in a system having an equilibrium configuration, $c_i^\dagger(\tilde{r})$ ıd $P_u^\dagger(\tilde{r})$, the inner potential $\varphi^\dagger(\tilde{r})$ depends on the charge density $\sigma^\dagger(\tilde{r})$ every surface [20], on the ionic concentration $c_i^\dagger(r)$ and on the pol- izability $\alpha^\dagger$ according to Eqs. (5), (21) and (22) of Reference 9a:

$$\varphi^\dagger(\tilde{r}') = \int \frac{\Sigma_i e_i c_i^\dagger}{|\tilde{r}-\tilde{r}'|}\, dV + \int \frac{\sigma^\dagger}{|\tilde{r}-\tilde{r}'|}\, dS - \int \alpha^\dagger \nabla\varphi^\dagger \cdot \nabla \frac{1}{|\tilde{r}-\tilde{r}'|}\, dV + \beta \qquad [33]$$

ıere β equals χ on the electrode and equals zero in solution and where † satisfies the Boltzmann relation:

$$c_i^\dagger = n_i^\dagger\, e^{-e_i\varphi_i^\dagger/kT} \Big/ \int e^{-e_i\varphi_i^\dagger/kT}\, dV \; . \qquad [34]$$

Introducing Eq. (22) of the present paper for $\tilde{P}_u$ into Eq. (5) of ference 9a for φ and into an analogous equation for φ* we also find

$$\cdot + m(\varphi^*-\varphi) = \int \frac{\Sigma_i e_i c_i}{|\tilde{r}-\tilde{r}'|}\, dV + \int \frac{\sigma^* + m(\sigma^*-\sigma)}{|\tilde{r}-\tilde{r}'|}\, dS - \int \alpha \nabla(\varphi^* - m\varphi^* - m\varphi) \cdot \nabla \frac{1}{|\tilde{r}-\tilde{r}'|}\, dV + \beta \; , \qquad [35]$$

ıere c_i is given by Eq. (23), $\alpha = \alpha_u + \alpha_e$, and the electrode charge den- ty is uniquely determined by the charges in solution and by the solvent larization.

We note from Eqs. (23) and (34) that $c_i^\dagger$ depends on $\varphi^\dagger$ in the same way that c_i depends on $\varphi^* + m(\varphi^*-\varphi)$ if we set $n_i^\dagger = n_i$. It then follows from Eqs. (33) and (35) that $\varphi^\dagger$ and $\varphi^* + m(\varphi^*-\varphi)$ satisfy the same integral equation if we set $\alpha^\dagger$ equal to α, (i.e., $D_s^\dagger$ equal to D_s), and if on each ionic surface we set $\sigma^\dagger$ equal to $\sigma^* + m(\sigma^*-\sigma)$. (There is no freedom of choice with the value of $\sigma^\dagger$ on the electrode surface, since it is uniquely determined by the magnitude and configuration of all the charges and dipoles in solution [9a].)

Since the potential function is unique, it follows that $\varphi^\dagger = \varphi^* + m(\varphi^*-\varphi)$, and accordingly, that $\underset{\sim}{E}^\dagger = \underset{\sim}{E}^* + m(\underset{\sim}{E}^*-\underset{\sim}{E})$, that $c_i^\dagger = c_i$ and $\underset{\sim}{P}_u^\dagger$ ($= \alpha_u \underset{\sim}{E}^\dagger$) equals α_u ($\underset{\sim}{E}^* + m\underset{\sim}{E}^* - m\underset{\sim}{E}$), and therefore (by Eq. (22)) equals $\underset{\sim}{P}$

A corollary of the relation between the σ's is:

$$e^\dagger = e^* + m(e^*-e) \quad , \qquad [35a]$$

where $e^\dagger$ is the charge on the central ion in the hypothetical system.

APPENDIX II. EVALUATION OF THE F_e OF AN EQUILIBRIUM SYSTEM

We shall have occasion to use an expression for the electrostatic free energy of an equilibrium system, F_e^{eq}. According to Eq. (43) of Reference 9a, it is given by

$$F_e^{eq} = \int_S \int_{\sigma=0}^{\sigma=\sigma} \varphi^{eq} d\sigma dS \quad , \qquad [36]$$

the integration being over all surfaces.

The value of φ^{eq} in an equilibrium system, consisting of a central ion of charge q, of an electrode M, of mobile ions, and of a medium of dielectric constant D_s, is [21]

$$\varphi^{eq} = \varphi_\rho^{\,q} + \frac{q}{D_s}\left(\frac{1}{r} - \frac{1}{r_i}\right) + \beta \quad , \qquad [37]$$

where r and r_i denote the distances from the field point to the center of the ion and to the latter's electrical image, respectively, and where φ_ρ is the contribution from the mobile ions, together with that form the electrode charges which they induce.

On the electrode surface, φ^{eq} is a constant, χ. Furthermore, on the surface of the central ion, $d\sigma$ equals $dq/(4\pi a^2)$ and also (it can be shown)

$$\int_{ion} \frac{dS}{r_i} = \frac{4\pi a^2}{R} \quad , \qquad [38]$$

where R is the distance from the center of the ion to its electrical im-

t $\varphi^* - \varphi + \beta$ is actually the potential which would be produced in an ilibrium system whose central ion had a charge (e*-e), whose volume rge density was zero, and whose polarizability was α_e. We see that its -value would be $\underset{\sim}{E}_v^* - \underset{\sim}{E}_v$. Using the relation which one obtains between $\underset{\sim}{E}_v$'s $\underset{\sim}{E}$'s in equilibrium systems when the usual ionic approximation is made], we find

$$\underset{\sim}{E}^* - \underset{\sim}{E} = (\underset{\sim}{E}_v^* - \underset{\sim}{E}_v)/(1 + 4\pi\alpha_e) \quad , \qquad [46]$$

 we note that the denominator equals D_{op}.

Combining Eqs. (22) and (46) we obtain

$$\underset{\sim}{P}_u = \alpha_u [\underset{\sim}{E}^* + m(\underset{\sim}{E}_v^* - \underset{\sim}{E}_v)/D_{op}] \quad . \qquad [47]$$

Introducing this value of $\underset{\sim}{P}_u$ into the general equation for the inner ential (Eq. (5) of Reference 9a) it can be shown that [24]

$$\varphi^* = \int \frac{\rho^*}{|\underset{\sim}{r} - \underset{\sim}{r}'|} dV + \int [\frac{\sigma^* - 4\pi\alpha_u m(\sigma^* - \sigma)/D_{op}}{|\underset{\sim}{r} - \underset{\sim}{r}'|}] dS$$

$$- \int (\alpha_e + \alpha_u) \nabla\varphi^* \cdot \nabla \frac{1}{|\underset{\sim}{r} - \underset{\sim}{r}'|} dV + \beta \quad . \qquad [49]$$

te: Eq. (48) is in Reference 24.]

Comparison with the equilibrium expression (33) shows that φ^* is also tually equal to the potential in an equilibrium system whose polariza- lity is $(\alpha_e + \alpha_u)$ and whose $\underset{\sim}{E}_v$-value is $[\underset{\sim}{E}_v^* - 4\pi\alpha_u m(\underset{\sim}{E}_v^* - \underset{\sim}{E}_v)/D_{op}]$. Using usual relation between $\underset{\sim}{E}_v$'s and $\underset{\sim}{E}$'s in equilibrium ionic systems], we find

$$\underset{\sim}{E}^*(r) = \frac{\underset{\sim}{E}_v^*}{D_s} - m(\underset{\sim}{E}_v^* - \underset{\sim}{E}_v)(\frac{1}{D_{op}} - \frac{1}{D_s}) \quad . \qquad [50]$$

s. (46) and (50) are the desired expressions for $\underset{\sim}{E}^*$ and $\underset{\sim}{E}$.

PENDIX VI. EVALUATION OF $F_e^* - F_e$ AND PROOF OF EQ. (26)

Introducing expressions (46), (47) and (50) of Appendix V into Eq. and into an analogous one for F_e it is found that

$$^* - F_e = \frac{1}{8\pi} \int [\frac{\underset{\sim}{E}_v^{*2} - \underset{\sim}{E}_v^2}{D_s} - (2m+1)(\underset{\sim}{E}_v^* - \underset{\sim}{E}_v)^2 (\frac{1}{D_{op}} - \frac{1}{D_s})] dV + (e - e^*)\chi \quad . \qquad [51]$$

This equation will be rewritten as:

$$F_e^* - F_e^A - F_e + F_e^B = -\frac{2m+1}{8\pi}(\frac{1}{D_{op}} - \frac{1}{D_s}) \int (\underset{\sim}{E}_v^* - \underset{\sim}{E}_v)^2 dV + T + w^* - w \qquad [52]$$

where T is defined by Eq. (53)

$$T = \frac{1}{8\pi D_s} \int (\underset{\sim}{E}_v{}^{*2} - \underset{\sim}{E}_v{}^2)\, dV - (w^* + F_e{}^A - w - F_e{}^B) + (e - e^*)\chi \quad . \qquad [53$$

Using Eq. (33) of Reference 9a, the integral in Eq. (52) can be written as [25]

$$\frac{1}{4\pi} \int (\underset{\sim}{E}_v{}^* - \underset{\sim}{E}_v) \cdot (\underset{\sim}{E}_v{}^* - \underset{\sim}{E}_v)\, dV = \frac{1}{4\pi} \int (\psi_v{}^* - \psi_v)(\sigma^* - \sigma)\, dS \quad . \qquad [54$$

According to Eqs. (16), (29), and (30) of Reference 9a, and to the rela tion $\rho^* = \rho$, we also have

$$\psi_v{}^* - \psi_v = (e^* - e)\left(\frac{1}{r} - \frac{1}{r_i}\right) \quad . \qquad [55$$

The integral in (54) is to be evaluated over the electrode surface and over that of the central ion. The electrode's contribution is zero since $\psi_v{}^* = \psi_v$ on the electrode according to Eq. (55) ($r = r_i$ there). The ion's contribution is readily evaluated with the aid of Eqs. (38) a (55). We find:

$$\frac{1}{4\pi} \int (\underset{\sim}{E}_v{}^* - \underset{\sim}{E}_v)^2\, dV = (\Delta e)^2\left(\frac{1}{a} - \frac{1}{R}\right) \quad . \qquad [56$$

We thus obtain

$$F_e{}^* - F_e - F_e{}^A + F_e{}^B = -\left(m + \frac{1}{2}\right)(\Delta e)^2\left(\frac{1}{a} - \frac{1}{R}\right)\left(\frac{1}{D_{op}} - \frac{1}{D_s}\right) + w^* - w + T \quad . \qquad [57$$

The LHS of Eq. (57) is simply $\Delta F^* - \Delta F$.

In Appendix VIII (a) it is inferred that to a good approximation

$$T = 0 \quad . \qquad [58$$

Using Eq. (13), Eq. (26) of the text is thereby established.

In utilizing Eq. (57) for evaluating $F_e{}^* - F_e$, we shall make use of the following expression for $F_e{}^B - F_e{}^A$, deduced from Eq. (41).

$$F_e{}^B - F_e{}^A = (kT \ln \gamma_B + e\varphi_S - e\chi + e^2/2a\ D_s) - (kT \ln \gamma_A + e^*\varphi_S - e^*\chi + e^{*2}/2a\ D_s) \qquad [59$$

APPENDIX VII. PROOF OF EQ. (25)

We obtain from Eqs. (44), (46), (56), (57) and (58)

$$\Delta F^* = w^\dagger - m(w^* - w) + \frac{m^2}{2}(\Delta e)^2\left(\frac{1}{a} - \frac{1}{R}\right)\left(\frac{1}{D_{op}} - \frac{1}{D_s}\right) + \frac{m(m+1)}{2D_s R}(\Delta e)^2 + kT \ln \gamma_\dagger \gamma_B{}^m/\gamma_A{}^{1+m}$$

$$- mT \quad . \qquad [60]$$

Appendix VIII (b) it is inferred that to a good approximation (the me as that made in deducing that T = 0),

$$w^{\dagger} - w^{*} - m(w^{*}-w) + \frac{m(m+1)}{2D_s R}(\Delta e)^2 + kT\, \ell n\, \gamma_{\dagger}\gamma_B{}^m/\gamma_A{}^{1+m} - mT = 0\,. \qquad [61]$$

. (25) of the text then follows from Eqs. (60) and (61).

PENDIX VIII. PROOF OF EQS. (58) AND (61)

) Proof of Eq. (58)

We evaluate separately the terms $\int(\underset{\sim}{E}_v{}^{*2}-\underset{\sim}{E}_v{}^2)\,dV/8\pi D_s$, $F_e{}^A+w^*$, and ^B+w that appear in Eq. (53) for T.

Setting $(\underset{\sim}{E}_v^2{}^{*}-\underset{\sim}{E}_v{}^2) = (\underset{\sim}{E}_v{}^{*}-\underset{\sim}{E}_v)\cdot(\underset{\sim}{E}_v{}^{*}+\underset{\sim}{E}_v)$, we obtain from Eq. (33) of ference 9a [27],

$$\frac{1}{8\pi D_s}\int(\underset{\sim}{E}_v{}^{*2}-\underset{\sim}{E}_v{}^2)\,dV = \frac{1}{2D_s}\int(\psi_v{}^{*}+\psi_v)(\sigma^{*}-\sigma)\,dS\quad . \qquad [62]$$

note that ψ_v is given [28] by Eq. (63) and that $\psi_v{}^{*}$ is given by the me equation if e is replaced by e*.

$$\psi_v = e\left(\frac{1}{r} - \frac{1}{r_i}\right) + D_s\varphi_{\rho\dagger}\quad . \qquad [63]$$

this equation $\varphi_{\rho\dagger}$ represents the potential due to the mobile ions and e electrode charges they induce, in the hypothetical system of Appendix

Both $\psi_v{}^{*}$ and ψ_v are zero on the electrode surface, since $(r-r_i)$ and $_\dagger$ vanish there (cf Reference 9a). Evaluating the integral in Eq. (62) er the surface of the central ion, we find

$$\frac{1}{8\pi D_s}\int(\underset{\sim}{E}_v{}^{*2}-\underset{\sim}{E}_v{}^2)\,dV = \frac{(e^{*2}-e^2)}{2D_s}\left(\frac{1}{a}-\frac{1}{R}\right) + \int_{\text{ion}}\varphi_{\rho\dagger}\,\frac{(e^{*}-e)}{4\pi a^2}\,dS\quad . \qquad [64]$$

Remembering that $F_e{}^B+w$ is the electrostatic free energy of an uilibrium system when the central ion B occupies the same position as the intermediate state, we can compute this quantity using Eq. (39). $^A+w^*$ can be computed in a similar way.

In this way we find for T

$$T = \int_{\text{ion}}\varphi_{\rho\dagger}\,\frac{(e^{*}-e)}{4\pi a^2}\,dS - \int_{\text{ion}}\int_{q=e}^{q=e^{*}}\frac{\varphi_\rho{}^q\,dq\,dS}{4\pi a^2}\quad . \qquad [65]$$

In the absence of added salt, these φ_ρ's are zero so that T is then actly zero. In the presence of added salt, an approximate solution

[29] of the Poisson-Boltzmann equation (solved in the presence of the central ion) leads to the relation

$$\varphi_\rho^{\,q}(\underset{\sim}{r}) = \varphi_{\rho^\dagger}(\underset{\sim}{r}) + (q-e^\dagger)\varphi_1(\underset{\sim}{r}) \quad , \qquad [66]$$

where $\varphi_1(\underset{\sim}{r})$ is independent of q.

Introducing this expression into Eq. (65) we obtain Eq. (67), where we have set $\varphi_1(r)$ equal to its value at the center of the ion, φ_1, say.

$$T = (m + \tfrac{1}{2})(\Delta e)^2\varphi_1 \quad . \qquad [67]$$

As noted above, φ_ρ and hence φ_1 equal zero when there is no added salt. If, in the presence of excess salt, the ion is effectively outside the double layer, these quantities can again be simply evaluated, utilizing the discussion following Eq. (39). Estimating φ_{atm} from the γ's, φ_1 is found to be negligible. In intermediate cases, too, we judge that the RHS of Eq. (67) is a small term [29], probably of the order of $(m+1/2)kT$ or less. Numerical computation shows that the term in Eq. (57 involving $(m+1/2)(\Delta e)^2/2a\ D_{op}$ is extremely large in comparison with $(m+1/2)kT$, and therefore we may set T = 0 in Eq. (57).

(b) Proof of Eq. (61)

Calculating $F_e^{\,B}$ and $F_e^{\,B}+w$ by the method described in Appendix II an subtracting we find

$$w = -\frac{e^2}{2D_sR} + \int_{ion}\int_{q=0}^{q=e}(\varphi_\rho^{\,q} - \varphi_S - \varphi_{atm}^{\,q})\frac{dq}{4\pi a^2}\,dS \quad . \qquad [68]$$

Analogous expressions can be written for w^* and $w^\dagger$, and we obtain for $w^\dagger - w^* - m(w^* - w)$,

$$w^\dagger - w^* - m(w^* - w) + \frac{m(m+1)(\Delta e)^2}{2D_sR} + kT\,\ln \gamma_\dagger\gamma_B^{\,m}/\gamma_A^{\,1+m} - mT = \int_{ion}\int_{q=e^*}^{q=e^\dagger}\frac{\varphi_\rho^{\,q}\,dq\,dS}{4\pi a^2}$$

$$- m\int_{ion}\varphi_{\rho^\dagger}\frac{(e^*-e)}{4\pi a^2}\,dS \quad . \qquad [69]$$

The RHS of Eq. (69) equals zero in the absence of added salt. In the presence of added salt we find with the aid of Eq. (66),

$$\text{LHS of Eq. (69)} = \frac{m^2}{2}(\Delta e)^2\varphi_1 \quad . \qquad [70]$$

This term can be neglected in comparison with the $m^2(\Delta e)^2/2a\ D_{op}$ which appears in Eq. (60), for reasons indicated in Part (a).

APPENDIX IX. MODIFICATION IN PROOF DUE TO PRESENCE OF FIXED, ADSORBED IONS IN THE ELECTRICAL DOUBLE LAYER

We consider here the modifications in the proofs of Appendices II to VIII which arise when the double layer contains the fixed, adsorbed ions discussed in the Refinements section of the text. The results deduced in Appendix I are unaffected.

In Eq. (37) of Appendix II there will be an additional contribution to the potential arising from each fixed ion of the electrical double layer. If q_k denotes the charge of the kth fixed ion and if r_k and r_k^i denote the distance from the kth ion and from its image to the field point, then the contribution to the potential arising from all fixed ions is, in an equilibrium polarization system,

$$\sum_k \frac{q_k}{D_s}\left(\frac{1}{r_k} - \frac{1}{r_k^i}\right) , \qquad [71]$$

the summation being over all fixed ions.

This can be shown to add the following terms to F_e^{eq} in Eq. (39):

$$\sum_k \int_{ion} \int_{q_k=0}^{q_k=q_k} \frac{\varphi_\rho^{q_k} dq_k}{4\pi a_k^2} dS + \sum_k \frac{q_k^2}{2D_s}\left(\frac{1}{a_k} - \frac{1}{R_k}\right) + \sum_{k\neq j}\sum \frac{q_k q_j}{2D_s}\left(\frac{1}{r_{jk}} - \frac{1}{R_{jk}}\right)$$

$$+ \sum_k \frac{q\, q_k}{D_s}\left(\frac{1}{\rho_k} - \frac{1}{\rho_k^i}\right) \qquad [72]$$

where a_k is the radius of the kth fixed ion; R_k, r_{jk}, R_{jk}, ρ_k and ρ_k^i denote the distance from this ion to its electrical image, to the jth fixed ion, to the latter's electrical image, to the central ion and to the latter's electrical image, respectively. The first term in Eq. (72) describes the interaction of the mobile ions with the fixed ions.

In Eqs. (40) and (41), φ_S now represents the potential in the body of the solution due to all the ions, fixed and mobile, of the double layer and so includes Eq. (71). Accordingly, $q\varphi_S$ in Eqs. (40) and (41) now includes the last term of (72). The remaining three terms of (72) should be added to Eqs. (40) and (41).

Eq. (42) of Appendix III relating F_e* and $F_e^\dagger$ is unaffected.

Eq. (43) of Appendix IV was deduced from Eq. (41). Because of a cancellation of the newly added terms, Eq. (43) remains unchanged (except for the new significance of φ_S noted above). Accordingly, Eq. (44) is also unchanged.

The equations in Appendix V are unaffected.

The newly added terms, given by (71), cancel in Eq. (55) of Appendix VI so that this equation and the remaining ones of this Appendix and of Appendix VII are unchanged.

In Appendix VIII, the terms (71) are added to Eq. (63), and this in turn causes the addition of certain terms to (64). The sum (72) is also added to the expressions for F_e^B+w and F_e^A+w* (with q equal to e and to e*, respectively). Once again, there is a cancellation and the final equations of Appendix VIII (a) and (b), (67) and (70) remain unchanged.

Accordingly, Eqs. (25) to (27) for ΔF* are unaltered.

APPENDIX X. EFFECTS OF A DEPENDENCE OF χ ON MEAN ELECTRODE CHARGE DENSIT

In the present paper, it has been assumed for simplicity that χ, the potential drop due to electrode-solvent and solvent-solvent interactions at the interface, was independent of the mean electrode charge density, $\bar{\bar{\sigma}}$ say. Recent work suggests, however, that the degree of orientation of the solvent molecules next to the electrode (and hence χ) may vary with this quantity [30]. In this Appendix we examine the effects of a dependence of χ on $\bar{\bar{\sigma}}$.

The term χ enters the various formulas in two ways: It appears in expressions for the potential in Eqs. (33), (35), (37) and (49), and in expressions for the free energy, Eqs. (39), (40), (41), (43), (44), (51), (53) and (59). In a previous paper [9a] we have considered the effects of treating χ as a function of $\bar{\bar{\sigma}}$ rather than as a constant. As discusse there, when a few electrons are transferred in some process, the average electrode charge density is affected only to a negligible extent. In such cases, χ is a constant for all states in the process. The argument made in the present paper using the expressions for the potential thus remain entirely unchanged. It was also noted [9a] that in the free energy expressions, a term, $\chi\int\sigma dS$, should be replaced for such processes by

$$\chi''(\bar{\bar{\sigma}}) + \chi'(\bar{\bar{\sigma}})\int\sigma dS \quad , \qquad [71]$$

where χ'' and χ' are constant for a given $\bar{\bar{\sigma}}$. It may be verified that thi substitution leaves the final equations, Eqs. (25) to (27), unchanged. The first term of (71) always cancels out in any equation deduced for a free energy change. Thus, in effect, χ is replaced by χ'. Since χ does not appear in the final equations, neither does χ'.

Accordingly, we infer that any effect of a change in degree of orientation in the solvent layer next to the electrode is a more indirect one. It might affect the "dielectric constant" in the vicinity of the electrode, particularly the contribution from orientation polarization. But, we observe from Eq. (26), unless D_s is close to D_{op}, changes in the former have very little effect on ΔF^*. Again, it might affect to some extent the distance of closest approach of the central ion. In some studies of the equilibrium properties of the electrode double layer, Grahame [30] has shown that a self-consistent interpretation of the data can be obtained assuming such an effect. Extremely interesting inferences were drawn about the behavior of the solvent in this region towards ions of different size. It is clear that an analogous study of the electrode kinetics of simple electron transfers at various electrode charge densities should be very interesting. At certain charge densities, it appeared from the equilibrium studies, there is no oriented solvent layer. The interpretation of kinetic data obtained under such conditions would be correspondingly simplified.

REFERENCES

1 cf reviews by (a) J. O'M. Bockris, Modern Aspects of Electrochemistry, Academic Press, Inc., New York, 1954, Chap. 4; (b) P. Delahay, New Instrumental Methods in Electrochemistry, Interscience Publishers, Inc., New York, 1954; (c) A. E. Remick, J. Chem. Ed., 33 (1956) 564.

2 cf J. E. B. Randles, Trans. Faraday Soc., 48(1952)828; A. A. Vlcek, Coll. Czech. Chem. Comm., 20(1955)894.

3 R. A. Marcus, Trans. N.Y. Acad. Sci., 19(1957)423.

4 R. A. Marcus, J. Chem. Phys., 24(1956)966).

5 R. A. Marcus, (a) J. Chem. Phys., 26(1957)867; (b) ibid., 26(1957) 872.

6 cf D. C. Grahame, Chem. Rev., 41(1947)441. It may be noted that when the double layer has such a small thickness, no penetration may be necessary for an electron transfer. The conditions cited in the text would then be fulfilled automatically.

7 cf J. Koutecky, Collection Czechoslov. Chem. Communs., 18(1953)183.

8 These conclusions can also be phrased [4] in terms of the application of the Franck-Condon principle to reaction (1). It was shown [4] that only when assumption (a) in the preceding section is valid can

this principle be applied to electron transfers in chemical (and electrochemical) reactions.

9 (a) R. A. Marcus, ONR Technical Report No. 11, Project No. NR 051-339, August 15, 1957; (b) cf J. Chem. Phys., 24(1956)979.

10 cf Reference 1a, Chap. 3, page 167.

11 Changes in electronic energy and entropy of the metal can be ignored in the path involving the minimum free energy of formation of X*. T low specific heat of metals shows that such changes correspond to large free energy increases.

12 Since then, a little more experimental information has become available on entropies of activation of electron transfer reactions in solution. These additional results tend to support the idea that p is approximately equal to its maximum value. These data will be dis cussed in a subsequent communication.

12a We shall use Eq. (31) to evaluate k_f. The probability that the ion will be in state X* and in a region of unit cross-sectional area in the interval R, R+dR, as compared with being in any unit volume, is $\exp(-\Delta F^*/kT)\,dR$. Multiplying this by $f(v_R)dv_R$, the probability that the ion has a velocity component perpendicular to the electrode in the range v_R, v_R+dv_R, we obtain the contribution to k_1/k_{-1} from this range dR dv_R. Multiplying this term by the corresponding value of k_{-1}, v_R/dR, and integrating over negative v_R's (motion away from electrode) we find

$$k_1 = \int_{-\infty}^{0} f(v_R)\,v_R\,dv_R\,e^{-\Delta F^*/kT} \quad .$$

Since $f(v_R)$ equals $\exp(-mv_R^2/2kT)/\int_{-\infty}^{+\infty}\exp(-mv_R{}^2/2kT)\,dv_R$, we obtain the formula given in the text upon integration.

13 As before, Eq. (31) is used. We first note that if the probability of reaction is very high, the probability that the ion will be in an particular interval (R,R+dR) depends on the recent history of the io In particular, it is a function of the probability that the ion has not undergone electron transfer shortly before reaching this dR. While such conditional probability considerations can play a role in the fast electron transfers in the gas phase, in solution the probability of reaction before reaching any dR is so small (because of ΔF*) that the recent history is essentially one of no reaction. In this case, the chance of finding the ion in a state X* in a region